Physics for the Rest of Us

Ten Basic Ideas of Twentieth-Century Physics That Everyone Should Know... and How They Have Shaped Our Culture and Consciousness

ROGER S. JONES

CB

CONTEMPORARY BOOKS

A TRIBUNE NEW MEDIA COMPANY

Library of Congress Cataloging-in-Publication Data

Jones, Roger S. (Roger Stanley), 1934–
    Physics for the rest of us ; ten basic ideas of twentieth-
century physics that everyone should know . . . and how they
have shaped our culture and consciousness / Roger S. Jones.
        p.   cm.
    Includes index.
    ISBN 0-8092-3939-6 (cloth)
    ISBN 0-8092-3716-4 (paper)
    1. Physics—Popular works.   I. Title.
QC24.5.J66   1992
539—dc20   92-20232
                                                           CIP

For Louise

Front cover photographs clockwise from upper left-hand corner:
Albert Einstein, FPG International; "Bucky Ball," Ken Eward/
Science Source; Lasers, Hank Morgan/VHSID Lab/ECE
Department UMA/Science Source; Mushroom cloud, FPG
International; Caffeine crystals, Dr. Jeremy Burgess/Science Photo
Library; and Andromeda galaxy, Tony Ward, Tetbury/Science Photo
Library

Copyright © 1992 by Roger S. Jones
All rights reserved
Published by Contemporary Books, Inc.
Two Prudential Plaza, Chicago, Illinois 60601-6790
Manufactured in the United States of America
International Standard Book Number: 0-8092-3939-6 (cloth)
                                                   0-8092-3716-4 (paper)
10   9   8   7   6   5

Contents

Acknowledgments

THIS BOOK HAS GROWN OUT of many years of reading, studying, teaching, conversation, and reflection, and there is no way to express my deep gratitude to the countless authors, friends, students, and colleagues who have given me so much precious advice and insight. But I must, at least, give heartfelt thanks to those who have helped me in shaping and expressing the ideas in this book:

To Larry Dossey and Richard Fuller for providing invaluable critiques on the manuscript,

To William Mishler for sharing his ideas with me over the years and for his sensitive reading of the manuscript,

To my dear friend Robert K. Anderson for his encouragement, inspiration, challenges, and unflagging support,

To Marion Rogers for her "miraculous intercession" and friendship,

To my agent, Elizabeth Backman, for her patience, diligence, and sage advice,

To my editor and publisher, Harvey Plotnick, for his acumen, his demanding critique of the book, and his faith in me,

And to my wife, Louise, to whom this book is dedicated, for her patient help, perennial encouragement, and sustaining love that made it all possible.

Und doch, 's will halt nicht gehn:
ich fühl's, und kann's nicht verstehen;
kann's nicht behalten,
doch auch nicht vergessen;
und fass' ich es ganz,
kann ich's nicht es messen!
—Richard Wagner
    Die Meistersinger von Nürnberg

And yet, it just won't go away:
I feel it and cannot understand it;
cannot hold on to it,
nor yet forget it;
and if I grasp it wholly,
I cannot measure it!

Introduction

TWENTIETH-CENTURY PHYSICS HAS GIVEN US atomic power, space travel, and computers. There's hardly an aspect of contemporary society or of our own individual lives that has not been profoundly transformed by the ideas and applications of relativity, cosmology, and quantum theory. The changes are most apparent in the technological marvels of engineering, electronics, communications, space travel, medicine, and weaponry that were undreamt of in the nineteenth century. Abraham Lincoln never had an x-ray, watched television, or used a word processor to write a speech.

The ideas of twentieth-century physics have had an even more dramatic influence on our deepest beliefs and assumptions about the universe we inhabit. The theory of relativity has not only unlocked the secret of atomic energy but has also released us from the bonds of rigid space and absolute time. The study of the heavens has given us not only the space program but a new story of cosmic creation as well. And while quantum theory has endowed us with microchips and superconductivity, it has also saddled us

1

with a strange and insubstantial view of the very stuff we are made of. The worldview we all share and take for granted today would have been inconceivable and incomprehensible to Lincoln or even to a nineteenth-century physicist.

A conceptual understanding of twentieth-century physics has become indispensable to any meaningful and thoughtful appreciation of the profound changes that have taken place in our contemporary spiritual beliefs as well as in our assumptions about the material world. Make no mistake about it, modern physics is as relevant to our notions of divinity as it is to our belief in atoms. In this book you will find ten basic ideas drawn from relativity, cosmology, and quantum theory that everyone should know in order to understand twentieth-century physics and to grasp how physics has shaped our culture and consciousness. My goal is to combine a conceptual approach to modern physics with an exploration of its deeper meaning and philosophical significance. Therefore, ten chapters of this book are devoted to the basic ideas of modern physics, and these ten chapters alternate with others that explore the ideas from humanistic, psychological, aesthetic, ethical, religious, and spiritual points of view. Along with an explication of the "basic text" of contemporary physics, a kind of running Talmudic commentary is presented in which questions are raised about the wider meaning and implications of physics:

- How does science affect our attitudes and beliefs about nature and the universe?
- Does science shed any light on the meaning and role of human existence?
- Are science and religion ultimately in conflict?
- Do aesthetic considerations affect scientific thought?
- What are the limitations and shortcomings of scientific theories?
- Are scientific theories independent of those who create them?

- What are the basic assumptions used in science and where do they come from?
- How has scientific thought changed and evolved over the ages?
- What are the ethical responsibilities of scientists?
- What criteria are appropriate in judging science and its consequences?
- Does science always provide the best and most appropriate methods for determining the truth?

In an attempt to deal with these questions, physics is approached here critically as well as humanistically. An essential reason for a nonscientist to study twentieth-century physics is so that he or she can judge and evaluate its meaning and consequences. The public is too easily persuaded that science is a matter exclusively for experts. It is not necessary to become an expert practitioner of science in order to judge it from a broadly humanistic and cultural point of view. One need not be an architect to judge the function and beauty of a building. Journalists commonly evaluate and report competently on political events although they are not statesmen or politicians. (To argue that science is an exception to other fields is to beg the whole question of scientific relevance and lay participation.) Expertise has its place in science as in other fields; but when it comes to the inevitable influences of science on society and human beliefs, the opinion and judgment of laypersons are fundamental. If physics intimidates, alienates, or frightens the rest of us, we shall never be able to understand and judge its values, uses, and dangers.

Why a special emphasis on the twentieth century? Modern physics, which includes among its cast of characters such exotic players as quarks, quasars, lasers, superconductors, black holes, pulsars, red giants, time warps, curved space, unified forces, superstrings, inflationary expansion, the big bang, and the big crunch, has had an immediate and pronounced effect on contemporary culture. But the subtle and often unconscious influence of modern physics on our thoughts and beliefs is even more powerful. When we watch "Star Trek," read an article on microchips, or listen to an account of some remote island galaxies of stars, some of us are unaware of the assumptions of relativistic space and time and of the quantum view of matter that underlie these presentations. Even our religious convictions about the origin and meaning of existence are affected by the latest theories of big bang evolution and the random, meaningless dance of atoms. What we don't know *can* hurt us.

The lack of a traditional background in the classical physics of the eighteenth and nineteenth centuries is no real barrier to your achieving a conceptual and qualitative understanding of modern physics. Concepts from classical physics can be developed as needed in the context of relativity and quantum theory, which is the approach used in this book. Certainly, scientists and engineers need a solid grounding in classical physics before embarking on their modern studies, because the mathematical theories and problem-solving techniques of the twentieth century are grounded in classical methods. (Incidentally, many engineers need very little modern physics at all.) But from the conceptual point of view, classical concepts often involve misconceptions and biases, which hamper our understanding of the radically new and unfamiliar ideas in quantum theory and relativity. The traditional assumption that one can learn modern physics only after a thorough study of classical physics simply does not apply to one's achieving a purely conceptual understanding of twentieth-century physics.

There is another important reason for exploring physics from a humanistic and critical point of view. Physics must be returned to its rightful place among the liberal arts. The artificial separation between the sciences and the humanities, which occurs both in educational institutions and in the world at large, has caused serious damage to our culture and civilization.

Science is regularly treated as something quite apart

from and outside of humanistic studies. It has grown a shell around itself that makes it impervious to the humanistic, aesthetic, and ethical analyses and critiques that are commonly applied to other fields. Because science is "objective" and deals only with the "truth" of the natural world, so the thinking goes, no criteria may be applied to it other than objectivity, empiricism, logic, consistency, and quantification. It stands apart from the general philosophical and humanistic background that is common to literature, art, history, and all other cultural studies. True, some scientists emphasize the importance of aesthetic considerations in the creations of their new theories, but ultimately these theories stand or fall not because of their beauty and meaning but because of their validity and utility.

As a result of this isolation of science from the humanities and its exemption from humanistic judgments, the meaning, implications, and practical consequences of science are never permitted to reflect back upon science itself. Science is a "pure," disinterested exploration of nature. What others do with the fruits of science may be subject to ethical and aesthetic judgment, but scientists are free to pursue their researches wherever they may lead.

We are now approaching the end of a century in which the fruits of science have been used to perpetrate mass genocide, to carry on the most destructive wars in history, to develop the most hideous weapons of annihilation, and to raise environmental pollution to an ever more perilous and irreversible level. As we approach the twenty-first century, can we still afford to leave judgments about science solely in the hands of the scientists themselves?

It is imperative today that we overcome the formidable barrier that the study of science presents to most nonscientists and laypersons—that is, to the rest of us. If society is ever to make humane, wise, and democratic choices about the future course of science and technology, the populace must have sufficient knowledge of science to think about and debate it intelligently. Scientific literacy has become a political as well as an intellectual necessity. Because professional technical training in science is neither possible nor even desirable for all of us, the introducing of alternative forms of science education and instruction—especially at the popular level—becomes crucial. A qualitative and conceptual approach is essential to improving physics literacy. Most popular presentations, however, fail to interpret physics, to analyze its meaning and implications, and to view it critically. Without analysis and criticism, we fail to recognize and evaluate the deep cultural and humanistic influences of science.

Any potential threat we face from science is not only from its technology but also from the power it has had to color our minds with a modern worldview that characteristically offers little nourishment or hope to the human spirit. The universe of twentieth-century physics is arbitrary, random, and meaningless—lacking in any human purpose or value. Contemporary science has had an inescapable and profound influence on our spiritual as well as material beliefs. Unless we expose and articulate the roots common to science, religion, art, and the humanities, we shall never learn how to judge the effects of science on our minds and hearts as well as our bodies. And we shall never learn how properly to guide the more humane development of science in the future. If physics is to serve and benefit humanity, then it must become physics for the rest of us.

Chapter 1

The Special Theory of Relativity

THERE ARE MYSTERIES IN THE NATURE of light that have eluded all the explanations of science. How does light originate and travel through space? How does it emerge from and enter into matter? What is light—substance, vibration, pure energy? Is color fundamental to light or merely a matter of perception? Science has found some partial answers to these persistent and challenging questions, but at the deepest level, they remain unanswered (and perhaps are unanswerable). Yet, the quest to understand light has given rise to the two great revolutions of twentieth-century physics—relativity and quantum theory.

WHAT IS LIGHT?

In seventeenth-century England a youthful Isaac Newton stood in a darkened room and directed a beam of sunlight against a glass prism. The white light was deflected and transformed into an iridescent apparition on a distant wall—a spectrum of rainbow colors. Newton pondered and further investigated this dispersion of light into its many hues. Ultimately, he concluded that white light is not something simple and indivisible. Rather, light is composed of color elements, or rays, which are separated by the prism and fanned out into a spectrum of component colors. Newton's analysis was controversial at first, but in the eighteenth century it was widely accepted and became influential. Still, it did not explain the actual nature of light: Was light some kind of wave spreading in a material medium, such as air or water, as Robert Hooke and Christian Huygens had believed; or did light consist of tiny particles, like projected pebbles or bullets, that may be bent or refracted in their passage through a prism, as Newton himself had conjectured?[1]

The controversy and speculations over light continued long after Newton's death, but during the course of the nineteenth century, the weight of the evidence gradually accumulated in support of light as a wave. For example, light was observed to spread out and bend as it passes the sharp edge of an opaque object, just as sound waves and water will bend around a corner. It was possible to interpret this *diffraction* of light, as it is called, only by assuming that light consists of waves and not of particles. In addition, the speed of light in transparent materials like glass and water was measured to be slower than the speed of light in the vacuum of empty space. The wave theory of light correctly predicted this, but the particle picture incorrectly predicted that light should travel more rapidly in glass than in empty space.

The final blow to the eighteenth-century particle picture of light was delivered by James Clerk Maxwell in his brilliant new theory of electromagnetism, published in 1873. Maxwell unified and described all knowledge of electrical and magnetic phenomena in terms of electromagnetic fields that could travel through space in the form of waves. Maxwell went on to show that light was one of

---

1. See J. Fauvel, R. Flood, M. Shortland, and R. Wilson, *Let Newton Be!* (New York: Oxford University Press, 1988), 91.

many possible kinds of electromagnetic waves, which today include radio and television waves, radar, microwaves, x-rays, and gamma rays. The dispersion of light into its component colors, as observed two hundred years earlier by Newton, as well as the diffraction and speed of light were completely and accurately described by Maxwell's new mathematical theory of light waves.

It must have seemed at the time that the nature of light had been determined once and for all. Indeed, there was toward the end of the nineteenth century a belief in the imminent completeness of science, as soon as the final details were attended to. Maxwell himself tried to quash the opinion abroad in his day that scientists soon would have nothing to do but alter the great physical constants by a few mere decimal places. Albert A. Michelson, a Nobel laureate in physics, wrote in 1898:

> While it is never safe to affirm that the future of Physical Science has no marvels in store even more astonishing than those of the past, it seems probable that most of the grand underlying principles have been firmly established and that further advances are to be sought chiefly in the rigorous application of these principles to all the phenomena which come under our notice.[2]

But it was neither the first nor the last time that prophets would mistakenly anticipate the "completeness" of science.

LIGHT AND RELATIVE MOTION

Among the details that required attention was one nagging consequence of Maxwell's theory. It predicted that the speed of light is always the same, whether measured by

---

2. See L. Badash, "The Completeness of Nineteenth-Century Science," *Isis*, vol. 63, no. 216 (March 1972): 48.

someone in motion or at rest. At first, this seems ridiculous. We have known how to measure and calculate relative speeds properly since the time of Galileo.

Suppose, for example, that I am standing at the side of a highway. I observe two cars approaching each other from opposite directions and measure their speeds as 40 mph and 30 mph, respectively. Now I ask myself, how fast is the 40-mph car moving according to the driver in the 30-mph car? Because each car is shortening, or "eating up," the gap between them—one at 30 mph and the other at 40 mph—that gap is closing at a rate of 70 mph. So the 30-mph driver sees the other car approach at a speed of 70 mph. (By the same token, the 40-mph driver clocks the other car at 70 mph.) To get the relative speed of each car with respect to the other, I have only to add the two speeds together.

Continuing in this vein, I go on to ask what the results would be if the two cars travel in the same direction. The 30-mph driver sees the other car pull ahead at 10 mph, and the 40-mph driver sees the other car fall behind at 10 mph. In this case, I have only to subtract the two speeds. So there it is: Galileo's simple recipe for calculating relative velocities. You add the speeds of approaching cars or subtract the speeds of cars moving in the same direction.

Maxwell, however, says that Galileo's recipe doesn't work for light. Take, for example, two cars—one moving toward you and one away. You observe the white headlights of the car moving toward you and the red taillights of the car moving away. According to Maxwell, you will measure the same speed for the white light of the approaching car and the red light of the receding car. The motion of the cars has no effect. According to Galileo, however, the two measurements would be different. The approaching car emits white light toward you. The white light is sped up by the car, and you must add the speed of the car and the light to get the speed you measure. The other car is moving away as it emits red light toward you. You must subtract the speed of the car from that of the light to get the slower speed of the red light. Maxwell, then, predicts the same speed, and Galileo predicts different speeds—a direct conflict.

How are we to settle such a conflict? The most common and persuasive approach in science is to look at the evidence. Empiricism, in fact, is at the heart of the so-called scientific method; and although there are other ways to argue for one scientific theory over another, as we shall see, we'll begin with tradition and examine the data.

An experiment was performed in 1926 that is very similar to the car example just discussed.[3] In this experiment, the speed of light was measured from a rotating pair of binary stars—one approaching the earth and the other receding from it.[4] The data showed unequivocally that the speed of light was the same from each star, thus vindicating Maxwell and demonstrating the incorrectness of the Galileo recipe.

Many other experiments have shown the invariance of the speed of light when measured in relative motion. The first experiments to test Maxwell's prediction were performed in the 1880s by two American physicists, Albert Michelson and Edward Morley. Because of his interest in the speed of light, Michelson had invented a special device called an interferometer, which used the wave properties of light to measure very small distances. He and Morley then carried out a series of experiments that could detect very small discrepancies in the speed of light under differing conditions of relative motion. All their experiments demonstrated that the speed of light was always the same in

3. See A. H. Jay and R. F. Sanford, *Astrophysical Journal*, vol. 69, no. 250 (1926).
4. Binary stars are pairs of stars that rotate about a common center. There are many such binary pairs in the skies. To perform this experiment, the observers simply chose one pair that is rotating in a plane that also includes the earth. In effect, you view the pair "edge on," so that one of the stars periodically approaches the earth while the other star recedes from it.

relative motion. Michelson later received the Nobel prize in physics for his pioneering work, and yet to the end of his life he was troubled by the results of his experiments.[5] He simply could not understand why there was no relative speed of light.

Like most of his contemporaries, Michelson believed that light behaved like sound waves or water waves that travel in a material medium. If, for example, you are rowing against the direction of water waves on a lake, you would measure the waves to be traveling faster than if you were stationary. In Michelson's day, most scientists believed that all space was permeated by a material medium that sustained light waves, just as air sustains sound waves. This hypothetical medium was called the *luminiferous aether*. Scientists tried to measure a relative speed of light as the earth moved through the aether, just as a rower would measure a relative speed of the waves while rowing against them. No relative speed for light was ever measured, and no evidence for the existence of the luminiferous aether was ever found. The aether theory was eventually laid to rest, and the Michelson-Morley experiments pounded the final nail in its coffin.

If Michelson had trouble believing his own results, then you can imagine how difficult it was for most physicists to accept them. Galileo's relative motion calculations, which had withstood two hundred years of testing and which agreed with all human experience, simply did not apply to light. No wonder Michelson doggedly continued to repeat his experiments over a period of decades.

CATCHING UP TO LIGHT

There was one physicist, however, who needed no convincing by Michelson or anyone else. He was Albert Einstein, who had independently convinced himself that you can

---

5. See A. Pais, *'Subtle Is the Lord . . .'* (New York: Oxford University Press, 1982), 114–15.

never "catch up" with a light beam.[6] At the age of sixteen, Einstein had contemplated a hypothetical experiment—a thought experiment, as he quaintly put it—in which he assumed for the sake of argument that Galileo was correct. Suppose, Einstein argued, I could attain the speed of light and then ride alongside a beam of light. According to Maxwell's theory, light is a vibrating electromagnetic wave. Therefore, if I move along at the speed of light, then I would see a wave that is stationary in space. This turns out to be an impossibility according to Maxwell, and contrary to experience as well. In Maxwell's theory, "stationary light" is impossible, and it has never been observed.

What Einstein meant is that a vibrating electromagnetic wave can never be at rest or stationary. The vibrations, or oscillations, of a wave are the result of both its motion and its spatial variation. When a spatially varying pattern moves past you, it appears to be changing in time. It's like seeing a movie. Each frame in the film is slightly different or varied from the one before it, and as the frames flash successively before your eyes, you perceive the effect of motion. If you stop the film (or move along the film at the same speed), the perception of motion disappears. Light is like the motion in a movie. You can "see" the motion (light), only when the film (light wave) is moving with respect to you. If you stop the film (wave), the motion (light) disappears.

Another way of stating the impossibility of catching up to light is to say that there is no *reference frame* in which light is at rest. A reference frame, which is a very basic idea in the theory of relativity, is simply the platform or framework from which one makes observations. It may be a laboratory or an observation deck, for example, and can include some experimental and observational devices—

---

6. The Michelson-Morley experiments seem not to have been crucial to the development of Einstein's thought—in fact, it is not known whether or not he knew about them. See G. Holton, "On the Origins of the Special Theory of Relativity," *Thematic Origins of Scientific Thought* (Cambridge: Harvard University Press, 1973), 165.

metersticks, clocks, voltmeters, telescopes, and so on. The reference frame may itself be in motion or at rest.

If you cannot reach the speed of light, Einstein argued, then you cannot go any part of the way toward the speed of light either; for otherwise you could leap from one reference frame to another, in which light would progressively go slower and slower until you caught up with light. Not only is this physically impossible, but it would require new contradictory laws of physics for certain reference frames in which light is "stationary." Thus, in order to avoid any contradictions in the laws of physics, Einstein postulated that the speed of light must be the same in all reference frames. This is a special case of the fundamental principle that the laws of physics cannot be different for observers in different reference frames.

THE PRINCIPLE OF RELATIVITY

The idea that the laws of physics must be the same for observers in different reference frames was, if anything, even better established and more fundamental in physics than Galileo's rules for combining relative velocities. This idea is called the principle of relativity, and it states, for example, that we may use the same laws of physics whether we design circuits for use on the earth or for a rocket to be sent to Jupiter.

Einstein had found a direct conflict between the principle of relativity and Galileo's rules for the relative speed of light. There was no question in Einstein's mind about which to reject. To preserve the logical consistency of the laws of physics as well as the validity of Maxwell's theory of electromagnetism, which correctly prohibits the possibility of "stationary" light, Galileo's rules had to go. Nothing less than the principle of relativity and the generality of physical law were at stake.

This was no minor decision. Two hundred years of experiment, observation, and applied physics had established the validity of Galileo's rules and of the vast framework of Newtonian mechanics that rested upon them. Countless phenomena, from the clockwork motion of the planets to the random behavior of the molecules in a gas, had been analyzed and explained with astonishing success. All of it had been accomplished with Newton's physics based on Galileo's laws of motion. How could Galileo possibly be wrong after such a fabulous track record?

Yet Einstein had convinced himself that the speed of light could not be relative—and, of course, he was right. The Michelson-Morley experiments in the 1880s, as well as the countless experimental verifications of the constancy of the speed of light that have been carried out since, testify to the uncanny soundness and accuracy of Einstein's intuition and logic. And so in the first years of the twentieth century, in his early twenties, Einstein set out resolutely to explore the inevitable consequences of the two ideas that he firmly believed must be right: the principle of relativity and the invariance of the speed of light. He took these two ideas as the fundamental postulates, or assumptions, of his revolutionary new special theory.

Consequences of the Postulates of Special Relativity

The special theory of relativity was first sprung upon an unsuspecting world in 1905 in the form of a now-classic and illustrious paper by Einstein.[7] The theory is referred to as "special" in contrast to the general theory of relativity, which was completed in 1915. The earlier theory deals only with the special case of motion at a constant speed along a straight line, such as a car moving at a steady fifty miles per hour along a very straight stretch of road. A car moving in this way—at a constant speed in a fixed direction—is said to have a constant velocity. When a car accelerates, its velocity changes, so a car moving at a constant velocity

7. A. Einstein with H. A. Lorentz, N. Minkowki, and H. Weyl, "On the Electrodynamics of Moving Bodies," *The Principle of Relativity* (New York: Dover, 1952, reprint of 1923 translation).

undergoes unaccelerated, or uniform, motion. The special theory of relativity deals only with such uniform motion at constant velocity, i.e., with the motion of objects moving on straight lines at constant speeds, such as a train moving at fifty miles per hour on a long straight track.

The restriction in the case of uniform motion is clearly artificial, and it was adopted by Einstein only to make his theory conceptually and mathematically simpler. Most motion is not uniform, but accelerated. Objects speed up while falling to the ground. Cars round turns and start and stop. Planets follow curved elliptical paths around the sun while going faster and slower at different points in their orbits. From the very beginning, Einstein recognized that the requirement of constant velocity would eventually have to go. The 1915 general theory removes this restriction and deals with all forms of motion and also with gravity, as we shall see in Chapter 3.

In the 1905 paper, Einstein confined himself to the simpler case of uniform motion. He proposed his two basic postulates of the special theory of relativity, which we shall paraphrase as follows:

1. The laws of physics are the same for observers in all uniformly moving reference frames.
2. The speed of light is the same for observers in all uniformly moving reference frames.

The first postulate had been accepted by all physicists already, essentially as a matter of faith. There was nothing new in it. The second postulate, however, was truly radical. The Michelson-Morley results were indeed puzzling and disturbing, but no one had imagined that light was an exception to the general rules for relative motion. Those rules were implicit in the first postulate: To compare the laws of physics in different reference frames, we must take account of their relative velocities. If I am observing moving bodies—billiard balls or electrons, for instance—in a rocket that is passing me, I must take the speed of the rocket into account. Galileo's rules for relative motion had always been assumed as a matter of course in comparing physical phenomena in moving reference frames.

Yet the rules for relative motion that are so basic to the first postulate don't apply to light, according to the second postulate. It remained for Einstein to demonstrate that the two postulates are indeed consistent, but only if we recognize that a new view of space and time is implied by the two postulates. In his paper, Einstein went on to develop and explore this new view of space and time, destroying forever the classical Newtonian concepts of absolute space and time.

SIMULTANEOUS EVENTS

Einstein began his paper by considering what his two postulates implied about motion (i.e., about the measurement of space and time). The idea of motion, which is central to all of physics, is described by using concepts such as speed and acceleration. Speed, for example, is a measure of how much distance is covered in a certain time. Therefore, in order to determine speed, it is necessary first to measure distance and time. The quantitative description of motion, which pervades all of physics, is based on the measurement of space and time.

The space-time consequences of Einstein's postulates are described in terms of three effects called the *relativity of simultaneity, time dilation,* and *length contraction.* Actually, they are all related and interconnected, but it is easier to understand them individually. We begin with the relativity of simultaneity, which demonstrates that simultaneity is a relative and not an absolute concept.

We shall again employ one of Einstein's famous thought experiments: Imagine one observer—Stacy, let's say—standing on the platform in a railroad station while a train is passing by. As luck would have it, two lightning bolts strike—one at the front end and one at the back end of the train—as it passes Stacy. Fortunately, no one is hurt, but the lightning leaves telltale char marks at both ends of the train and also at two corresponding points on the platform. Stacy takes advantage of the situation to make an observation. She notices that the light from the two bolts reached her at the same instant. She then paces off the distance between her point of observation and each of the char marks on the platform and discovers that the two distances are equal. In other words, Stacy was exactly equidistant from the two ends of the passing train as the lightning bolts struck. What does Stacy conclude?

Stacy received the light from the two lightning events at the same instant. The distances between Stacy and the two events are equal, and the speed at which light traveled to her from the two events is, of course, the same. So the two light signals Stacy received traveled equal distances in equal times and reached her at the same instant. The two original events (the emission of the light by the two bolts) must have occurred at the same time, and therefore they were simultaneous.

There seems to be nothing too surprising in all of this. As far as Stacy is concerned, two bolts stuck at equal distances from her, and since she received the light signals at the same instant, the bolts must have struck simultaneously.

Let's now look at the same two events from the point of view of an observer in the reference frame of the moving train. Our second observer will be Trent, who is located at the center of the train. Because Trent is moving with the train, he is approaching the light signal that travels toward him from the front end of the train. He's also moving away from the light signal that proceeds toward him from the back end of the train. Thus Trent first intercepts the front-end light signal, while the back-end light signal is still catching up with him. An instant later in time, he receives the back-end signal.[8]

---

8. The time delay between these signals will admittedly be very small because of the high speed of light. But we can imagine that Trent is

Afterwards, Trent also paces off the distance between his point of observation and the char marks at each end of the train. He, too, finds that the two distances are equal.[9] What then does Trent conclude? He is equidistant from the two events, and the speed of light is the same for him as it is for Stacy. (Here is where the crucial second postulate comes in: the speed of light is exactly the same in the train and in the station reference frames.) The two light signals traveled equal distances to Trent in equal times. But the light from the front event reached him earlier than the light from the rear event. Therefore the front event must have occurred *earlier* than the rear event.

The two events that are simultaneous for Stacy are not simultaneous for Trent. Stacy claims that the two bolts struck at the same instant; Trent says they occurred at different times. Who is right?

Einstein's surprising answer is that they are both right. The simultaneity of two events is not an absolute or invariant notion for all observers. The time delay between two events is different in different moving reference frames. Simultaneity is a relative concept, not an absolute one.

You may object and argue that Stacy must be right because she was at rest with respect to the moving light signals while Trent was not. The conclusion in the train reference frame must be wrong, you might say, because Trent did not properly take into account his relative motion with respect to the light signals. But this argument ignores the first postulate, which states that the laws of physics are just as valid in Trent's reference frame as in Stacy's.

It is the combination of the two postulates that reveals the full radical significance of relativity. The laws of phys-

---

also moving at a very high speed—half or three-quarters of the speed of light. In any case, there is a time delay, no matter how small, as long as Trent is moving.

9. The equality of the two distances that Stacy and Trent measure is not critical to the argument; it's just a convenience. They can infer the time intervals from whatever distances they measure. But since this is a thought experiment, we might as well keep things simple.

ics *and* the speed of light are the same in both reference frames. Relativity is a most democratic theory: one reference frame (the station) is no better than any other (the train). Trent has as much right as does Stacy to make his own measurements and to draw his own equally valid conclusions, which happen to be different from Stacy's—different, but equally correct. There's no escaping it: They are each correct, Stacy in her reference frame and Trent in his. In fact, using Einstein's theory, each of them can figure out what conclusion the other observer will reach, and it's all perfectly consistent. They have different but equally valid ways of describing the same two events. It's like one person using centimeters and another using inches to describe the length of a table. Both descriptions are correct. There is no contradiction once you relinquish the incorrect notion of absolute simultaneity. The simultaneity of two events is a characteristic of a particular reference frame. It is not a universal or absolute description.

SPACE AND TIME

The relativity of simultaneity is not only startling in itself but is also interconnected with two other remarkable space-time effects called *time dilation* and *length contraction*. In fact, the three effects are one and the same. They are just different aspects of the revolutionary new description of space and time that follows from the postulates of special relativity. Once the relativity of simultaneity has been established, time dilation and length contraction follow as inevitably as night follows day.

Time dilation means that an earth observer will see the clock on a moving rocket ship recording time at a reduced rate (as if the seconds on the moving clock were dilated or stretched out). Suppose that the earth and rocket clocks are synchronized at noon, just as the rocket passes the earth traveling at a high speed. An hour later when the earth clock reads 1:00 P.M., the earth observer will read 12:45 P.M. on the rocket clock. An hour on the earth will correspond to only forty-five minutes on the rocket. Just as the simultaneity of two events is not absolute but depends on the reference frame in which the two events are being observed, so the measured interval of time between two events also depends on the frame of reference.

If that isn't bad enough, then consider this. When the rocket observer reads 1:00 P.M. on the rocket clock, she will read 12:45 P.M. on the earth clock. Each observer will see the other's clock slowed down. It's a reciprocal effect and works both ways.[10]

Then there is length contraction. The earth observer will measure the length of the moving rocket to be shorter, or contracted, as compared with its length when at rest. If the rocket observer measures the length of the rocket at 100 yards (remember that to the rocket observer, the rocket is at rest), then the earth observer will measure the moving rocket's length at 75 yards. (And vice versa for two rockets that "pass in the night.")

All of this relativity of space and time might seem incredible, but actually it's quite logical and, in fact, necessary once we realize that the speed of light is the same for all observers. After all, speed is determined from the measured values of distance divided by time. If the speed of light is always the same regardless of the motion of different reference frames, then clearly distance and time must have different values in different frames. The only way to keep distance divided by time (speed) the same is to allow both the distance and the time to vary. For the sake of argument, let's imagine that the speed of light is 2 mph. In one reference frame we measure that light travels 6 miles in

---

10. You may wonder which observer is right: what happens when the rocket returns and the clocks are compared? Special relativity cannot properly answer this question because it deals only with uniform, unaccelerated motion. The rocket cannot turn around and return to earth without decelerating and accelerating again. As far as special relativity is concerned, the two clocks will forever disagree as they continue to recede from each other. There is an answer, however, to this paradox (known as the twin's paradox), which will be discussed later in this chapter.

3 hours, so we determine a speed of $\frac{6 \text{ miles}}{3 \text{ hours}} = 2$ mph. In another frame the corresponding distance and time are measured as 8 miles and 4 hours, giving a speed of $\frac{8 \text{ miles}}{4 \text{ hours}} =$ 2 mph. If the speed of light is invariant for all observers, then space and time must vary.

Simultaneity, time intervals, and length must all be relative. Two events that are observed to be simultaneous in one reference frame will not be simultaneous in any other frame that is moving with respect to the first frame. A moving clock will be observed to record time more slowly than a stationary clock. A meterstick moving past you will not measure one meter long but something less, depending on its speed. And as strange as these effects may seem, they have all been verified in countless experiments, especially, but not exclusively, involving atoms and elementary particles.

At this point, you may well protest that all of this is fine and good for physicists. Perhaps the speed of light is invariant, and space and time are relative, and these things may be very logical and consistent in physics. But, you may ask, why have I never seen any evidence of it? I travel in cars and on jets, and I've never seen moving clocks slow down or lengths shorten. Besides, how can we possibly make time schedules for jet travel if all these clocks are slowing down (or is that why planes are always late)? And even if scientists can see all of these things now, why did they never see them before Einstein?

I can hardly blame anyone for being incredulous. These are perfectly reasonable questions and they deserve reasonable answers. As a matter of fact, Einstein himself never observed any of these effects before he came up with his theory. The reason why neither he, nor anyone before him, nor most of us since have ever seen these effects is that their size is insignificant at the speeds typical of normal human experience. In order to see a rocket shorten to three-quarters of its normal length at rest, the rocket would have to travel at two-thirds the speed of light, or about 450 million miles an hour! Even to see a rocket's length decrease by 1 percent would require it to travel at a speed of 94 million miles an hour. The highest speeds rockets can reach today are a thousand times slower than this. At ordinary jet speeds the changes in time and length are only a few parts in a trillion. Such effects are completely unobservable and undetectable without extraordinarily accurate measuring devices.[11] The following table summarizes how the effects of time dilation and length contraction change with increasing speed. At low speeds, the effects are extremely small, but as speeds approach that of light, the effects become very large and dramatic.

LENGTH-TIME CONTRACTION-DILATION TABLE

Speed (mph)	Fraction of Speed of Light	Factor of Contraction or Dilation	100 Yards Contracts to (yd.)	1 Hour Dilates to (hr.:min.:sec.)
670	0.000001	1.0000000000005	100	1:00:00
67,000	0.0001	1.000000005	100	1:00:00
6,700,000	0.01	1.00005	99.99	1:00:02
67,000,000	0.1	1.005	99.5	1:00:18
168,000,000	0.25	1.0328	96.8	1:02
336,000,000	0.5	1.1547	86.6	1:09
503,000,000	0.75	1.5119	66.1	1:31
604,000,000	0.9	2.294	43.6	2:18
664,000,000	0.99	7.089	14.1	7:05
670,000,000	0.999	22.37	4.47	22:22
671,000,000	0.99999	224	0.45	223:36

In the microrealm of elementary-particle physics, we can accelerate electrons and protons to speeds just a snail's pace slower than that of light. In that realm, these relativistic space-time effects are commonplace. In fact, the giant accelerators and atom smashers used in particle-physics research would never work if Einstein's laws had not been taken into account.

Even the high-speed cosmic rays that bombard the

---

11. For an account of measuring time dilation in ordinary jet travel, see Chapter 3 in C. M. Will, *Was Einstein Right?* (New York: Basic Books, 1986).

upper layers of the earth's atmosphere provide strong evidence of time dilation. In the upper atmosphere, cosmic rays produce a certain particle—the muon—that can reach us at the earth's surface only because of the slowing of the moving muon's internal clock, which ticks off its short, unstable lifetime. At rest, muons live only two-millionths of a second on the average—a time far too short for them to travel through the 150 miles or so of the earth's atmosphere. The muons, however, travel so fast that their clocks go several hundred times slower than stationary earth clocks, which enables the muons to reach the earth's surface—and they reach it in great abundance.

Relativistic space-time effects lead to other unbelievable predictions. For example, suppose one twin takes off from the earth in a very high-speed rocket and leaves her twin brother behind. The rocket travels at very nearly the speed of light and makes a round-trip to a nearby star in the course of twenty years of earth time. When the traveling twin returns to her brother, who has aged twenty years, she will have aged perhaps by only a year and a half (the amount of the discrepancy will depend on the rocket's speed). The time intervals measured by the earthbound and the traveling twin will result in a permanent age difference between them. As strange and implausible as this sounds, it has been experimentally verified, not with people but with "twin clocks."[12]

As noted earlier, this effect really goes beyond the validity of the special theory of relativity. One twin must decelerate and accelerate again to turn around and make the round-trip. But special relativity deals only with unaccelerated motion at constant velocity. The problem can be analyzed, however, in the general theory of relativity (which will be discussed in Chapter 3). What the general theory shows is that there is a physical difference, or asymmetry, between the traveling twin in an accelerated refer-

12. The discrepancy was measured between a traveling and an earthbound clock that are otherwise identical, as described in the reference in footnote 11.

ence frame and the earthbound twin in a "stationary" reference frame (stationary as compared with the very high rocket speed). The physical acceleration of the rocket frame is not simply a matter of relative motion, and it causes the rocket time to be permanently delayed behind earth time.

RELATIVITY AND INVARIANCE

The relativity of simultaneity, time, and space may seem to suggest that anything goes and that there's no longer any natural order in the physical world. In fact, Einstein's theory is often simplistically interpreted to mean that "everything is relative." In popular presentations, relativity often implies a kind of chaos in which nothing is fixed, dependable, or permanent, and one person's rules (or one society's) are as good as another's.

Nothing could be further from the truth of Einstein's theory of relativity. For, although the measurement of space and time are different for different observers, those differences are neither arbitrary nor random. They follow a mathematically prescribed pattern that reveals a deep order and invariance underlying the prescribed differences. Space and time may vary individually from reference frame to reference frame, but the combined space-time separation between two events never does: it is an invariant.[13] Furthermore, relative motion may be different for different observers, but the speed of light is the same for all. It, too, is an invariant. The laws that prescribe how to translate one observer's space-time measurements into another's are also invariant and independent of the vagaries of relative motion. In fact, relativity is replete with invariant and conserved quantities.

Rather than suggesting that everything is relative, Einstein's theory demonstrates a deep and unchanging

---

13. More precisely, although space and time individually are different in different reference frames, a certain combination of the measurements of space and time, called the *invariant interval*, is the same in all reference frames and provides a bedrock of permanence underlying all the changes. The invariant interval will be discussed in Chapter 2.

order that underlies natural phenomena. Early misconceptions about relativity prompted Einstein to suggest a new name—invariance theory—to describe the deep permanence beneath the appearances of change. But "relativity" had caught on in both the scientific and popular imagination, and it was too late, even for Einstein, to change the name of his theory.

Unfortunately, misunderstandings and distortions of the meaning of relativity have continued to plague the theory. But despite its name, Einstein's theory is not about chaos and disorder but about harmony and invariance.

MATERIAL CONSEQUENCES

The space-time consequences of the postulates of the special theory of relativity are really only the beginning of the story. Einstein went on to consider how material bodies and their motions are affected by the new space-time ideas. In physics, to describe the behavior and activity of material bodies, concepts such as kinetic energy—energy of motion—and momentum are used. These concepts are defined by combining space-time notions with the material properties of a body. Momentum, for example, is defined as the product of the velocity and the mass of an object. Velocity, of course, involves both distance and time, while mass is a measure of the amount of matter a body possesses (2 grams, for instance, or 150 kilograms). So the definition of momentum and the physical laws governing it must change as a result of the new concepts of space and time, on which they depend.

When Einstein put all of this together in the effort to reformulate the laws of motion of material bodies (thus forever revising Newton's mechanics), he found something quite unexpected: The mass and the energy of a body were no longer independent quantities as in Newtonian physics. Each could be converted into the other—matter into energy and energy into matter.

Up until the time of Einstein, it had been believed that energy and matter were separately and independently conserved, for instance, in a chemical reaction. When hydrogen and oxygen combine to form water, the number of grams of water is equal to the number of grams of the original hydrogen and oxygen. Heat energy may be given off in the reaction, but this was thought to come from the energy stored in the original atoms, so that the mass and the energy in the chemical process could separately be accounted for. Einstein showed that the heat energy released actually depletes a tiny quantity of the mass of the original atoms. The mass of a water molecule is slightly less than the mass of the original hydrogen and oxygen atoms. It's very difficult to measure this tiny depletion in most chemical reactions, but in the far more powerful reactions involving the nucleus of the atom, the conversion of mass into energy is readily measurable.

It's as if matter is a concentrated form of stored-up energy. No longer did physicists and chemists believe that matter and energy were separately conserved in independent conservation laws, as they all had thought up to the time of Einstein. The combined mass-energy content of a body (or group of interacting bodies) is conserved in one single conservation law. And mass is now known to be just another form of energy, along with chemical energy, nuclear energy, heat, kinetic energy, and electrical energy.

All of this is described in Einstein's well-known equation (undoubtedly the most famous equation in the world), $E = mc^2$, which tells us how much energy we can get from a given quantity of matter. If you multiply the mass of an object by the speed of light squared, you find out how much energy can be produced by converting all the matter in the body into pure energy. Let's suppose, for example, that you could convert one gram of matter completely into energy. Now, one gram isn't a great deal of matter. A penny has a mass of about three grams. So imagine you could convert one-third of a penny completely into heat energy. How much water do you imagine you could boil?[14]

_____

14. To be a little more precise, how much water could you raise from the freezing point at 0°C (32°F) to the boiling point at 100°C (or 212°F)?

What would you guess—a pint, a gallon, a hundred gallons, a thousand gallons . . . ?

The answer is roughly forty-three million gallons, or an amount of water that would fill about eighty-five Olympic swimming pools!

The enormous amount of energy that can be released in the conversion of matter is the key to atomic energy and to the fantastic destructive power of the hydrogen bomb. These, too, are legacies of the consequences of the postulates of special relativity.

TRUTH OR CONSEQUENCES

The special theory of relativity is clearly no mere curiosity or collection of abstract and hypothetical ideas. Its consequences are all too real and practical. And yet it originated in the thought experiments of Albert Einstein's mind. It was motivated by a faith in the ultimate consistency and logic of physical law and by Einstein's desire to preserve that consistency, even at the cost of losing the well-substantiated and time-honored physics of Isaac Newton. Einstein was driven to give up the cherished notions of absolute space and time. True, these were assumptions built into Newtonian physics, but few scientists or philosophers ever thought of them as assumptions—as mental conventions or conveniences, constructed to buttress the magnificent edifice of the Newtonian world.[15]

A faith in consistency, order, and harmony is motivated as much by an aesthetic sense as by a quest for truth. Do the relativistic descriptions of space, time, and matter constitute true statements about the physical world? Or are they merely consequences of certain convenient assumptions—aesthetic choices—with only a limited and temporary utility for human endeavor? Do space and time themselves really have an existence independent of the human

---

15. Both Newton and Einstein, however, were acutely and explicitly aware of these assumptions.

mind that contemplates, conceives of, and describes them? The canons of absolute space, time, and reality, long associated (whether rightly or wrongly) with Newtonian physics, have forever been discredited and rejected by relativity. In Chapter 2 we shall ponder some of the questions that the relativity revolution has raised about space, time, and reality and their relationship to the human mind.

CHAPTER 2

Metaphors of Space and Time, Part I

THE SPECIAL THEORY OF RELATIVITY has forever transformed our ideas about the nature of space and time. Einstein no longer assumed, as had Newton, that space is fixed
and "rigid" or that time "flows" at the same uniform rate
for everyone. As we have seen, special relativity does not
even treat space and time as separate and distinct from each
other. Einstein has shown us that one observer's space is
another observer's time! In fact, there has been an evolution of space-time concepts in modern physics. But what is
actually evolving—our notions about space and time, or
space and time themselves? Can we actually distinguish the
"physical" or "objective" character of space and time from
our conceptions and inferences about them? In this chapter
we shall explore where our ideas about space and time
come from and why and how we choose one view over
another.

THE SPACE-TIME CONTINUUM

"Here time becomes space." These prophetic words, written by Richard Wagner in 1877, are sung in his opera

Parsifal by the sage knight Gurnemanz as he is about to initiate the opera's hero into the mysteries of the Knights of the Grail. Wagner could hardly have known when he wrote these words that they were to take on a literal meaning that would characterize a new era in science within thirty years.

In the railroad example of the previous chapter, we considered two lightning-bolt events that were judged to be simultaneous by an observer in the platform reference frame. The same two events, however, were determined not to be simultaneous in the moving-train frame. For Stacy, on the platform, the two events occurred at exactly the same instant and were separated simply by the length of the train. For Trent, on the train, one event occurred before the other, and so the two events were separated by an interval of time as well as by an interval of space—the length of the train. What was purely a spatial separation, or length interval, for Stacy was both a temporal and a spatial interval for Trent.[1] Stacy's space became Trent's space-time.

Furthermore, it isn't only the time measurements that are different in the two reference frames. The length of the moving train as measured by Stacy also differs from Trent's measurement. (The length of the moving train is contracted for Stacy.) But although the two observers measure different times and lengths, there is something about the events that relativity tells us is the same for both observers. A certain mathematical quantity, called the *invariant interval*, is defined in relativity. Just as velocity and acceleration are defined from space and time, the invariant interval is also defined mathematically from the prior measurements of space and time. But the invariant interval has a special and peculiar property, which makes it identical for Stacy and Trent. In fact, the invariant interval is identical for

---

1. Note that an event is defined both by its location in space and by the time of its occurrence. In relativity, the position and time of an event is different in different reference frames, and so the distance between the two events, as well as the interval of time between them, will vary from reference frame to reference frame.

anyone observing the same two lightning events in any other reference frame. The invariant interval represents a separation between two events that is the same for all uniformly moving observers.

The common use of the invariant interval in the theory of relativity suggests a completely new conception of space and time. For that which is invariant, or the same, for observers in different reference frames is neither length nor time separately, as in the Newtonian worldview. Rather, it is a certain mathematical interval that combines space and time together into one new concept—*space-time*. In the space-time approach, time is treated as a fourth dimension to be added to our three-dimensional space.

REPRESENTING SPACE-TIME

Imagine a long road in the desert, as straight as an arrow. A car travels on the road—now starting and stopping, now accelerating and decelerating, now moving at a constant speed, but always going in the same forward direction. Let's imagine the desert to be perfectly flat without any hills or valleys, and let's assume the road to be perfectly straight without deviations or curves. Then we can think of this as an example of motion purely in one dimension by ignoring the flat expanse of the surrounding desert and concentrating only on the straight, one-dimensional path of the car. The car does not move about freely on a two-dimensional surface as a ship does on the sea, and certainly it does not move in three-dimensional space as does a submarine or a space rocket. The car is limited to the one dimension of the straight, flat road. If we draw the car's motion on a map, it will be a straight line—a one-dimensional representation. This representation, however, does not portray motion as such but only the spatial path of the car. To fully represent the motion of the car, we need to indicate time as well. We often do this by showing both

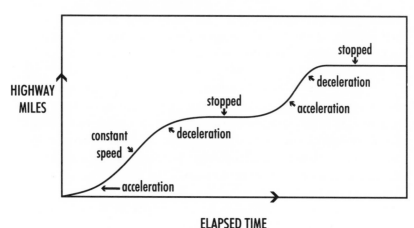

ELAPSED TIME

distance and time on a single graph, which then provides a better picture of the car's motion.

The snaking curve on this graph of distance against time represents the motion of the car in time. Where the line curves, it represents acceleration—speeding up—or deceleration—slowing down. Where it slopes upward at a constant angle, the car is moving at a constant speed. Where it is horizontal, the car has stopped.

Regardless of the details and specific techniques for interpreting such a motion graph, one thing about it is clear. The curve representing the motion is not drawn in one dimension but in two. On the graph, time is represented symbolically as a *spatial* dimension along the horizontal axis of the graph, and highway distance is represented on the vertical axis. The graph represents the sequence of the car's locations in space, i.e., miles along the road, as they occur in time.

If we watch a movie of the car shot from a helicopter high above the highway, we would see a small dot moving along a straight line. The path of the dot would appear straight, even though it speeds up and slows down at different times. If we inspect the film itself, each frame would show the dot in only one position, and we would lose all sense of time. In fact, we know that the sense of motion and change in a movie is an illusion, in part because of the persistence of vision, which creates an impression of continuous motion.

By contrast, the graph represents the whole "story" of the motion, all at once. It shows all the successive positions or points occupied by the car at each instant during its journey. It is not simply a portrayal of the car's motion but a representation of its total history, as if "frozen" in space-time. Although we do not experience time as a spatial dimension, we can represent it that way, and doing so gives us a picture of the unfolding (or rather, unfolded) story of the car. It's as if you laid out a row of photographs of yourself at sequential stages in your life. You'd have a kind of representation of your life story. But for our simpler story of the car's journey, we can see the parallel and coordinate roles played by space and time, formed into a two-dimensional continuum—space-time. We have a space-time picture of the entire story of the car's journey.

To see how all of this applies to physical space and relativity, imagine now the motion of some object in three-dimensional space—a rocket journeying to Jupiter, for example. In order to portray the story of this journey, you would have to add to the three dimensions of interplanetary space a fourth dimension of time (just as we added a second dimension of time to the one-dimensional path of the car). We cannot draw a "graph" to represent this journey as we could with the car because we do not live in a physical space of four dimensions, nor can we imagine what it would look like.[2] But, in principle, we can conceive of such a representation (which can, in fact, be described very adequately, although abstractly, in mathematical terms). The four-dimensional representation of the story of a moving body is really what we mean by space-time, and space-time, thus conceived, has been built into the mathematics and graphical representations of the theory of relativity.

2. For an excellent introduction to the romance of dimensions, see E. A. Abbott, *Flatland* (New York: Dover, 1952).

SPACE, ANYONE?

In working with relativity, scientists regularly use a four-dimensional space-time description of physical events, whether of lightning in a railroad station, electrons in an atom, or galaxies in an expanding universe. No one actually sees or visualizes such spaces, but they are so useful and common in scientific work that we tend to forget their representational and constructed character and to think about space-time as though it were as real as physical space.

As real as physical space! Come to think of it, how do we know that physical space is real? You might argue that unlike four-dimensional space-time, the three-dimensional physical space in which we live and breathe can be seen. Is that really true? Can we actually see space? In reality, we cannot. What happens is more like this: Our eyes receive information in the form of light, which our minds organize and interpret as images of objects in space. Space itself, whatever it may be, is invisible. We see only images of material bodies, which we organize according to a plan or pattern that we interpret as three-dimensional space. Our perception of space is actually a construct of the mind. Because we can never truly separate our perception of space from the physical reality of space (whatever that is), our experience of space is basically a construct of the mind as well.

One of the three dimensions—depth—is clearly a matter of mental interpretation. Most of us have binocular vision, which enables the brain to coordinate the slightly different information received by our two eyes and to interpret it as depth or distance between us and an object. Depth perception (as you can easily discover by viewing things with one eye closed) is constructed by the mind out of information from binocular vision. All the tricks and illusions of 3-D pictures and holograms depend on this fact.

As to the other two dimensions of so-called physical

space, they, too, are constructed from a varied pattern of light, dark, and color that fills the field of our vision. Light of varying intensity and quality (interpreted by the mind as degrees of brightness and hues of color) strike the retina. This pattern is perceived by the mind as objects located in an external space. But "seeing" objects and space is the end result of a complex process involving optics, biochemistry, and psychology. First, light is received through the optical mechanism of the eye. The light that strikes the retina induces nerve impulses, which are carried along the optic nerve to the brain. These impulses, in turn, are "interpreted" in the optical centers of the brain as the "physical" space around us.

You do not actually see a car moving along a highway or a tree outside of your window. What you see is a colored blob (a car) changing against a stationary background or some green and brown blobs (a tree) surrounded by various other shapes and blobs. Learning to distinguish cars and trees from the rest of our visual field is a matter of interpretation and training. Children learn to do it without knowing it, of course. People who were born blind and have their vision restored as adults often never succeed in doing it. They may never overcome the blooming confusion of an unfamiliar visual world and learn to construct a surrounding space out of it.[3]

No one knows better than magicians and illusionists how easy it is to make people see what is not there. We regularly acknowledge that what we see in mirrors, through lenses, and in photographs aren't "real" objects but mere images—mirages. We can usually distinguish a photograph or mirror image from "reality" by certain visual clues. If these clues are ambiguous or deliberately concealed (as in a magic show or in a carnival house of mirrors), we can become quite confused about image and reality. The greatest of all illusionists, though, is the hu-

---

3. See R. Gregory, *Eye and Brain: On the Psychology of Seeing* (New York: McGraw Hill, 1970).

man mind itself. We often fail to recognize (or, at least, to recall) that the vision of our eyes is also an image, which we interpret with our minds. Everything we see is a matter of interpretation and mental construction. We can never completely separate the sources of our images (whatever they may be) from the images themselves and our interpretations of them. The eighteenth-century German philosopher Immanuel Kant has gone so far as to claim that space is a pure, pre-existing form of human sensibility, or sense intuition, that conditions the mind to understand the voluminous, confusing sensory data we constantly receive.

The problems we encounter in trying to imagine a space of four dimensions pale in comparison with the tricks our minds play in getting us to see the three we take for granted. We may envision the space of Newton or Einstein, or we may picture spaces that are geometrical, astronomical, geological, subterranean, anatomical—or purely imaginary, mythical, or astrological. In all such cases, we are making a mental construct through a complex act of human imagination, however unconscious and automatic it may be. The spaces, not only of science but of everyday life in which human beings imagine themselves living and dying, are creations of the human mind, just like the metaphors of poets and artists.

CONSTRUCTING SPACE

Surely there is a difference between all these constructions, you may argue. Physical space is not arbitrary or purely imaginary. We never see most of the terrain on the ocean floor, but we can reconstruct and map it from sonar and radar readings, which do not depend on the eyes' images and the mind's constructs. Similarly, the anatomical space of the body can be reconstructed nowadays from CAT scans. Surely, these are no acts of imagination, but objective facts.

But are they? These radar and x-ray techniques are all modern versions of *sounding*, or plumbing the depths, which is still done on ships navigating in shallow waters. Instead of lowering a weight to find the floor of the sea, we now use a variety of ultrasonic and electromagnetic waves that bounce off objects. In the case of CAT scans, these waves intercept body organs. From such complex data, a picture must be constructed. In this day and age, such constructions are routinely made by computers, and, obviously, the computer constructions are programmed by human beings and are based on their preconceived notions about the structure and character of space. The data must still be interpreted by the mind.

The whole process is not all that different from what a blind man implicitly does in constructing a "picture" of space from the tapping of his cane. Initially, in learning to use a cane, a blind man is acutely aware of the sensations of the handle of the cane against the palm of his hand. Gradually, he learns to transfer his awareness to the tip of the cane. He imagines he can feel the tip touching the floor and the objects around him, although clearly this is a mental interpretation, based on the actual physical sensations in his palm. Ultimately, he even loses the awareness of the cane tip and constructs a "picture" (however crude) of the space around him from his complex sensory data as interpreted by his mind. He comes "to see" the space of his environment.[4]

Those of us who are sighted, in fact, employ the same process, but instead of a cane striking against the palm, we use light rays striking the retina. Like the experienced blind man, we have no awareness of the physical sensations at all, and deal only with the automatic end product—a mental image of space. Whether we are sounding, CAT scanning, tapping, or seeing, we construct space imaginatively—we create a metaphor. Because we can never fully separate the image from whatever the reality may be, what

---

4. The process of blind "seeing" and its relation to tacit knowledge is discussed in more detail in Chapter 1 of M. Polanyi, *The Tacit Dimension* (Garden City, N.Y.: Doubleday, 1967).

we truly experience is the world of our constructs and metaphors.

EVOLVING SPACE-TIME

From the scientific point of view, our metaphors of space and time are models or theories that we continually test against the evidence. But, as we've seen, that testing is also a matter of interpretation, and our models can change as new information becomes available and new theories of nature arise. The Newtonian model of the world was the culmination of a long process that had gradually displaced older Greek and medieval conceptions of the cosmos. And, in turn, at the opening of the twentieth century, Einstein's relative space-time eclipsed and replaced Newton's absolute space and time.

Greek space was ordered spherically in a hierarchy starting with the element of earth at the center, then moving outward to water, air, and fire to complete the sublunary realm, and finally to the celestial realm of the stars and planets.[5] Greek space also had a kind of organic mental character that is completely unfamiliar to us. For Plato, thoughts, feelings, and perceptions were forms of motion and change in space, and there was much less distinction than we make between the mental and the physical.[6] Platonic space is as much a mental realm as a physical one.

This mixture of the mental and the physical can still be found in medieval notions of space.[7] There is no better illustration of this than medieval astrology, in which mental experiences, sensations, and emotions are inextricably bound up with astronomical events in physical space. It seems strange to us—this marriage of the psychological and the physical—but it fits in perfectly with the organic,

5. G. F. Parker, *A Short Account of Greek Philosophy* (New York: Harper and Row, 1967), 145–46.
6. O. Barfield, *Saving the Appearances* (London: Faber & Faber, 1957), 101.
7. Ibid., Chapter 11.

participatory, and symbolic character of medieval space metaphors.[8] Indeed, part of the incongruity for us is that we have moved through history several stages beyond the space metaphors of medieval Europe and ancient Greece, and we are no longer aware of them. In fact, much of the ridiculing and categorical rejection of astrology by modern scientists is because scientists have a kind of blind faith in the current metaphors of space and time and are ignorant of former ones.

METAPHORS AND MEANING

If our conception of space and time (or space-time, as the case may be) is not a simple reflection of an objective physical space, then what else is it? Kant's explanation, as we've seen, is that space is a pure form of sensibility that allows the mind to order and rationalize raw-sense data. The mind must presort raw data into acceptable and meaningful categories, such as space, time, and causality, according to Kant. Imagine how chaotic our experience of the world would be if we could not order and systematize all the information collected by our eyes—if we could not *put things in their place*. Indeed, space is precisely about putting things in place.

Space is a metaphor of location and of relationship as well; for to make sense of things, we must not only place and locate them but also understand the connections among them. Space provides us with a scheme of locations and with a web of relationships—orientation, direction (up and down, left and right, in and out), distance (near and far), connection, separation, distinctness, identity. The very notion of existence is itself intimately tied to space. To exist, according to its Latin root, is *to stand out*. And what does one stand out from? From nothing other than space. Space is the background from which we stand out and become articulated as separate and distinct entities. It is the

8. R. S. Jones, *Physics as Metaphor* (Minneapolis: University of Minnesota Press, 1990), 63–70.

platform of our identity. It is the stage on which all action, interaction, and association is played out. It is the metaphor par excellence of relationship.

And yet, our modern space—our metaphor of relationship—has been emptied of many of its relational characteristics. We no longer conceive of space as organic, hierarchical, and participatory, nor as a realm of mind, knowledge, and wisdom, as did our Greek and medieval forebears. We envision our space as empty, void—the absence of everything. If we believe that Greek and medieval conceptions of space reflected the mental constructs and preconceptions of earlier minds, then isn't it just possible that our modern empty and sterile space might likewise reflect our own preconceptions and beliefs? If so, what are we to infer from such a sterile and inhospitable environment? Is it a suitable home for human occupancy? What are we trying to tell ourselves?

There are indications that we may be filling space once again with meaning and substance, as in earlier times. In Einstein's general theory of relativity, as we shall see in Chapter 3, space-time becomes equated with matter and energy themselves, and even with gravity. In modern quantum theory it is the empty vacuum of space itself that is seen as the source of all the matter and energy in the physical universe. In fact, it is mere vacuum that is conjectured to have given birth to the big bang in which our cosmos originated.

Have we then reversed our evolutionary march toward the austere emptiness of a featureless void? Are we becoming more conscious of the nourishment and succor that our existence in space requires for both our physical and mental well-being? And are our metaphors beginning to reveal their human origins?

CHAPTER 3

The General Theory of Relativity

FROM THE OUTSET, EINSTEIN KNEW that his special theory of relativity was artificially limited and had to be extended or generalized. The special theory deals only with reference frames in motion at constant speeds—uniform motion, as Newton called it. The laws of physics, however, must be valid in any reference frame regardless of its motion. When a rocket speeds up, or accelerates, the circuits on board don't suddenly stop working until the rocket resumes uniform motion at a higher constant speed. Furthermore, the special theory took no account of the force of gravity, which is present throughout the universe in the form of gravitational fields of varying strength. We are held on the earth by its gravitational field, the earth and planets are kept in their orbits by the sun's gravitational field, the stars of the Milky Way are held together in a galactic group by their mutual gravitational attraction, and ultimately the behavior of the entire cosmos is governed by gravity. Einstein also suspected that the problems of acceleration and gravity were not independent and would be solved together. Soon after the 1905 publication of the special theory, Einstein set out to fathom the mysteries of gravity. His quest for a general theory of relativity would take ten years and would culminate in the most extraordinary epiphany of his life.

THE ELEVATOR EXPERIMENT

The key to generalizing the earlier theory of relativity involves another of Einstein's famous thought experiments. In the effort to understand the connection between gravity and acceleration, Einstein imagined himself in an elevator. He wondered why he was standing on the floor. It seems like an obvious question—even a little naive—and yet the answer conceals one of the best-kept secrets in science. Normally, we would say that Einstein stands on the floor because the earth's gravitational attraction holds him there. If he drops a key from his pocket, it, too, will fall to the floor because gravity pulls it down.

Let's eavesdrop on Einstein's thinking. "All well and good," Einstein reasons, "but if I can't see out of the elevator, how do I really know I'm on the earth? After all, there's another way to explain why I'm standing on the floor. My elevator could be in outer space, hitched to the back of a rocket ship that is accelerating 'upward.'[1] As the elevator accelerates upward, I am pressed 'downward' toward the floor, just as happens in an elevator on the earth when it begins to accelerate upward in a building. And in my accelerated elevator, if I release a key from my hand, it, too, will 'drop.'

"As soon as I let go of the key, it stops accelerating, but it continues to move at the speed it had when I released it. This is just the law of inertia in operation: Bodies always move at constant speed unless some force accelerates them. And in empty space, far from any planets or stars, there is

---

1. *Up* and *down* have no real meaning in space, far from the earth. The term *upward* is used here as a convenient way of referring to the direction toward the roof of the elevator—the direction in which the elevator is being accelerated by the rocket.

no significant force of gravity or anything else to accelerate the key once I let go of it.

"Suppose the elevator is moving at 1,000 mph when I release the key. A second later, I am moving at 1,001 mph because I am still accelerating, but the key continues to move at 1,000 mph. The elevator, with me in it, continues to accelerate to 1,002 mph, 1,003 mph, 1,004 mph, and so on, but the key, once released, no longer accelerates. It keeps moving at the same 1,000 mph, while I go faster and faster. The key then 'falls behind' because the elevator floor 'catches up' with the key. Eventually, the floor reaches the key, but to me, it simply appears as if the key 'fell' to the floor.[2] What's more, if the elevator accelerates at just the right rate, the key will appear to fall at the same rate as if it were on the earth.

"Whether I am in an elevator at rest on the earth's surface or in an accelerated elevator in outer space, I see and experience exactly the same thing. In each case, I am held firmly against the floor, and the key I release falls to the floor at the same rate. Unless I can see outside my elevator, I cannot tell whether I'm in the earth's gravitational field or accelerating in outer space.

"Now, of course, if the rocket stops and starts or decelerates and accelerates, then I and the objects inside the elevator would move differently, and I could easily tell that I'm not on the earth. But if the rocket accelerates smoothly and at a constant rate, then no experiment I perform or observation I make can distinguish between the two cases. From an empirical point of view, they are equivalent."

The Principle of Equivalence

Through thought experiments like this one, Einstein discovered a remarkable equivalence between gravity and ac-

---

2. Visualizing all this relative motion may be a little easier if you imagine yourself at a stationary position in space watching Einstein and the key inside a glass-walled elevator. The elevator accelerates, but the key moves at a constant speed. So the key "falls behind" inside the elevator.

celeration. He later described this realization, which first occurred to him in 1907, as "the happiest thought of my life." This profound connection between gravity and acceleration became the cornerstone of the general theory of relativity. Einstein called it the *principle of equivalence*. It states that a uniformly accelerated reference frame is equivalent to a stationary reference frame in a uniform gravitational field (like the uniform field that exists over a small portion of the earth's surface, which is essentially flat[3]). More loosely, it says that an accelerated reference frame is equivalent to a gravitational field.

The equivalence principle also makes explicit a fact of nature that, though previously known, was rarely noted before Einstein. (It's the "best-kept secret" referred to earlier.) The famous, though probably apocryphal, legend of Galileo dropping an iron ball and a wooden ball at the same time from the Leaning Tower of Pisa hints at the equivalence principle: All bodies fall at the same rate in a gravitational field. We aren't often aware of this, because things we see falling, such as leaves or feathers, are impeded in their fall by the resistance of the air. But in a vacuum, or with dense compact objects like Galileo's iron and wooden balls, it is possible to verify that different bodies fall at the same rate in the earth's gravitational field.

Consider an experiment that Galileo couldn't do but which Einstein's argument suggests. Suppose you release

---

3. Just as a magnetic field is described by its strength (it's stronger the closer you are to the magnet) and its direction (a compass needle points in different directions at different locations on the earth), so a gravitational field is also described by its strength and direction. The earth's gravitational field loses strength as you rise above the earth's surface. Furthermore, because falling bodies move directly toward the center of the earth, the direction of the earth's gravitational field changes as you move around the globe. A uniform gravitational field is one that does not vary in strength or direction from place to place. The earth's gravitational field is uniform to a very high degree over a region of several square miles of the earth's surface. This is so because over such a region, the curvature of the earth is very slight, and there is no significant variation in the direction of falling bodies. Also, the strength of the earth's gravitational field does not vary appreciably from one location to another on the earth's surface.

an iron ball and a wooden ball simultaneously inside a rocket ship that is accelerating in space. Just as with our earlier falling-key experiment, the two balls would continue to move at the speed they had at the instant you released them, while you and the rocket would accelerate. Thus the two balls would appear to you to "fall" at the same rate. You could not determine whether you were in an accelerated rocket in space or in a stationary rocket in a gravitational field. To all intents and purposes, the two cases would be equivalent.

While Galileo had observed and stated the empirical fact that all objects fall at the same rate in the earth's gravitational field (apart from air resistance), Einstein went beyond Galileo and *interpreted* and explained this observation in terms of acceleration rather than as some mysterious force of gravity. This is a completely new way of understanding the force of gravity—in terms of the motion of one's reference frame—and it is the seminal idea of the general theory of relativity. In Einstein's universe, the force of gravity would become superfluous. The problem was whether the equivalence principle could be generalized to explain all gravitational fields. Einstein had his work cut out for him.

This peculiar equivalence principle is unique to gravitational fields—it does not apply to other fields. In an electric field, for instance, all electrically charged bodies do not accelerate or "fall" at the same rate. And in a magnetic field, different pieces of iron fall at different rates. But a planet, a comet, or a particle of dust, at equal distances from the sun, will all fall at the same rate in the sun's gravitational field. It is because of this unique property of gravitational fields that Einstein's thought experiment works. Electrically charged objects would all appear to fall at the same rate in Einstein's accelerated elevator, but at different rates in an electric field. So there is no difficulty in distinguishing acceleration from electric or magnetic fields. Uniform acceleration and gravity, however, cannot be distinguished; they are equivalent.

ACCELERATION AND REFERENCE FRAMES

The concept of acceleration is central to Einstein's general theory of relativity, which goes beyond the special case of uniform motion to the more general case of accelerated motion. We need to examine exactly what acceleration means in order to understand how Einstein constructed his new theory.

Basically acceleration refers to a change in either the rate or the direction of motion. If you are in a falling elevator, for example, you are moving at an increasing rate. You are in an accelerated reference frame. If you look outside as you pass by another elevator that is descending at a *constant* speed, it will appear to you to be accelerating upward because of your own downward acceleration. The appearance of a moving object depends on your own state of motion. If you are accelerating (i.e., in an accelerated reference frame), a body at rest or moving uniformly at constant speed will appear accelerated to you.

This is fairly apparent when only your rate of speed is changing, but what about the other aspect of acceleration— a change in the direction of motion? Why does acceleration involve changes in direction as well as in speed?

Let's suppose, for example, that you are riding on a merry-go-round. Your direction of motion is continually changing as you move in a circle. You look at a car that is moving along a straight road at constant speed. To you, because of your constantly changing direction, the car will appear to move on a curved path. Your motion on the rotating merry-go-round will make uniform motion look curved *relative to you.*

"Fine," you may say, "uniform motion appears curved, but what has that to do with acceleration?"

Accelerated motion includes changes in direction because the definition of acceleration in physics refers to *any* departure from motion along a straight line at constant speed. You may protest that if you move along a curve but do not speed up or slow down, then you are not accelerating. But curved motion is also a form of acceleration, which we can understand by examining the experience of motion. You can easily recognize acceleration if you are in a jet that is speeding up. You feel yourself pressed back into the seat. If the jet slows down, you feel yourself lurch forward. But if the jet moves smoothly forward, even at 500 mph, you can't even tell you're moving. Your body can sense acceleration, but it cannot detect uniform motion.

Suppose the jet banks to the right on a curved path. You feel yourself leaning to the left. Once again, your body tells you that you are not in uniform motion—you are not moving in a straight line at constant speed. In either case— changing speed or moving on a curve—your body senses acceleration. In physics, the definition of accelerated motion in effect formalizes our bodily sensations. Any departure from straight-line motion at constant speed is acceleration, and that means moving on a curve as well as changing speed. If you are in a reference frame that is not moving uniformly—a falling elevator, an accelerating car, a merry-go-round, or a roller coaster—then you are in an accelerated reference frame.

INERTIAL REFERENCE FRAMES

Our discussion of acceleration and accelerated reference frames makes clear that we cannot judge motion merely by its appearance. Whether a body appears to us to be in uniform or accelerated motion depends on the motion of our reference frame. How then do we ever know when something is truly at rest or moving uniformly? Einstein was very much aware of this problem, and so he devised a surefire way to determine whether a reference frame is moving uniformly. Once we establish that our reference frame is moving uniformly, then we can trust our observations of moving bodies. To establish this, Einstein used the law of inertia.

Descartes had first stated the law of inertia to describe the natural motion of an isolated body, i.e., a body upon which no forces, such as friction or gravity, for example, are acting. According to the law of inertia, such an isolated body moves at a constant speed along a straight line—uniformly, in Newton's language. Newton's whole system of mechanics—the study of the motion of material bodies—is a grand scheme to explain *departures* from the law of inertia: What happens when forces *do* act on bodies? How does their motion differ from their natural uniform motion, as described by the law of inertia? Departures from natural uniform motion always appear as acceleration and are explained by Newton in terms of the action of forces. In effect, Newton distinguishes between two kinds of motion—natural, or inertial, motion that is uniform, and force-induced motion that is accelerated.

The law of inertia can be observed readily in a laboratory, under the right conditions. A block sliding on a table normally slows down because a force of friction retards its motion. If we can reduce or eliminate friction, the block will slide without slowing down until it strikes something. A hockey puck slides quite far on ice because the friction between the puck and the ice is very low. We can reduce friction even further by "floating" a block on forced air or using magnetic fields and thus verify the law of inertia with very high accuracy. In televised broadcasts of space exploration—from inside a space capsule, for example—we also see direct evidence of the law of inertia as astronauts toss around objects, which move along straight lines at constant speed.

But if you are in an accelerated reference frame, you must be very careful, as we have seen. Genuinely uniform motion will appear accelerated (changing in speed or direction), if you observe it from a falling elevator or a merry-go-round. When you see an accelerated body, you cannot be sure whether its motion is truly accelerated or the result of your own accelerated motion.

To overcome this problem, Einstein proposed an empirical test, using the law of inertia. If you observe a moving object on which no forces are acting and that object patently obeys the law of inertia by moving uniformly, then you can be certain that your reference frame is not accelerated and must itself be moving uniformly. Since true uniform motion, according to the law of inertia, occurs when no forces act on a body, then if you observe a force-free body moving uniformly, your reference frame must also be moving uniformly with respect to that body.

This is the acid test that Einstein devised to distinguish uniformly moving reference frames from accelerated ones. Any reference frame in which isolated bodies (bodies on which no forces act) move according to the law of inertia is called an inertial reference frame. This is a much more precise and verifiable test to apply to reference frames than the looser, though admittedly more intuitive, idea of uniform motion. To check for the uniform motion of your reference frame, you must measure its speed with respect to some other reference frame, which must in turn be tested for uniform motion with respect to yet another reference frame. There's no end to this process; but you can test the law of inertia within your own reference frame, without referring to any other reference frame. So we have a practical and unambiguous definition of an inertial reference frame, which from now on we shall refer to as an IRF. We shall also call a reference frame undergoing any arbitrary kind of nonuniform motion simply an RF.

In effect, the special theory of relativity is about space, time, matter, and energy in IRFs. It states that the laws of physics and the speed of light are the same in all IRFs, but not in all RFs. Einstein knew, however, that the laws of physics must be the same in any RF. In the general theory, he sought to remove the limitations on the special theory and to reformulate the laws of physics so that they would be the same in any RF, regardless of its motion.

ACCELERATION AND CURVATURE

We can now return to the principle of equivalence, which links the whole problem of accelerated motion to gravity.

Einstein realized that he could not solve the problem of
acceleration unless he dealt simultaneously with gravity.
What he didn't yet suspect was that the relationship between gravity and acceleration would ultimately enlarge
and blossom into a radically new conception of the universe in which the very idea of gravitational attraction
among bodies would disappear along with Newton's distinction between inertial and forced motion.

If we observe inertial motion, we can easily distinguish between IRFs and RFs, but what if we observe accelerated motion? The principle of equivalence establishes an
ambiguous connection between uniform or inertial motion and accelerated motion by showing that the presence
of a gravitational field makes an IRF appear to be an
accelerated RF. You cannot distinguish between an IRF in
the earth's gravitational field and an accelerated RF.

In an IRF, we may see accelerating bodies changing
speed and following curved paths. In an accelerated RF, we
can see much the same thing. From a roller-coaster car, for.
instance, objects may appear to be accelerated because of
the relative motion of the car. The car speeds up and slows
down, climbs over curved hills, and swoops down into
rounded valleys. The relative motion of the car makes
bodies at rest on the ground appear to change their speed
and move on curves. The observation of accelerating objects blurs the distinction between IRFs and RFs.

At this point, Einstein made a revealing discovery
about acceleration and the curvature of motion. If we examine how acceleration is represented or portrayed, we find
that *all* accelerated motion can be characterized as curved,
even that of a car speeding up on a straight highway. This
becomes apparent when we use space-time to picture motion. In Chapter 2, we began the discussion of space-time
by graphing distance against time for a car moving on a
straight highway (see the Elapsed Time/Highway Miles
illustration in Chapter 2). In the course of its journey in
Chapter 2, the car underwent several kinds of motion—
speeding up, slowing down, moving at constant speed, and

stopping. Viewed from an airplane above the highway, the path of the car appears straight. But on the two-dimensional space-time graph, the car's motion has portions that are straight and portions that are curved.

In a two-dimensional space-time picture, uniform motion is represented by a straight line, while accelerated motion is represented by a curved line. To understand this better, we'll examine a two-dimensional space-time graph of two cases—an elevator descending at a constant speed and an elevator falling freely in the earth's gravitational field.

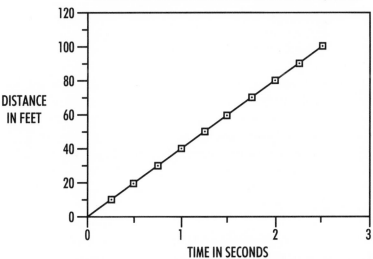

ELEVATOR DESCENDING AT CONSTANT SPEED

When an elevator is descending at a constant speed, the plot is a straight line because the elevator covers equal 10-foot intervals of distance in equal quarter-second intervals of time. This is really what is meant by uniform, or constant, speed. When an elevator falls freely (and therefore accelerates), it moves through greater distances in each successive interval of time. In the first quarter-second, the elevator moves 1 foot. In the second quarter-second, the elevator moves 3 feet. In the third quarter-second, the ele-

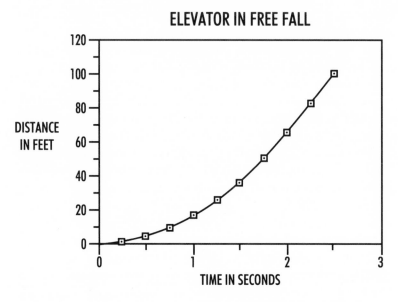

vator moves 5 feet, and so on. In the last quarter-second on the graph, the elevator moves 19 feet. The line on the graph curves upward to indicate the increasing distance the accelerated elevator covers in each successive quarter-second interval. Now we see that uniform motion appears straight on a space-time picture, and accelerated motion appears curved. In space-time, accelerated motion is curved, and uniform motion is straight. As a matter of fact, all accelerated motion, when plotted against time, exhibits curvature. In the four-dimensional world of space-time, an accelerated body always follows a curved path.

WHO NEEDS GRAVITY!

We return now to Einstein, who is ready to deliver the coup de grace. "Very well, then," he continues. "In my accelerated roller-coaster RF, objects that are at rest or moving uniformly along the ground will *appear* to me to be accelerated. If I plot their motion as I see it on a space-time graph, it will be curved. On the other hand, if I plot the motion of these bodies as seen in an IRF (at rest on the ground, for example), the space-time picture of their uniform inertial motion will be straight. For an observer in an IRF, the space-time picture of inertial motion is straight. For an observer in an accelerated RF, however, the space-time picture of inertial motion is curved.

"When I observe accelerated motion, all bets are off. If my IRF is in a gravitational field, I will again see curved motion. After all, if I throw a ball on the earth, it will follow a curved path. And whether I make a space-time plot of a falling elevator or of a pitched ball, the picture will be curved, just as if I were observing motion from an accelerated RF. Therefore, space-time curvature—the hallmark of acceleration—cannot distinguish between an accelerated RF and an IRF in a gravitational field. If that test fails, and I can devise no other foolproof way to distinguish between an IRF and other RFs, then what am I left with?

"What I am left with is nothing less than a new theory of gravity!

"Look here. Let's turn the tables around. If no sound empirical test can distinguish between a gravitational IRF and an accelerated RF, then perhaps the fault lies elsewhere. Perhaps there is no meaningful distinction between the two cases. If gravity and acceleration are equivalent and indistinguishable, then why do I need both of them?

"In fact, we don't need gravity at all! No one has ever seen the force of gravity. We see projectiles curving in the air and planets revolving about the sun. We *infer* a force of gravity from these curved accelerated motions, but we never actually see or otherwise observe such a force. All our knowledge of the force of gravity is indirect—an inference from the motion it reputedly causes. If all we have of the force of gravity is the curved motion it produces, and if that same curved motion also characterizes accelerated RFs, then why complicate life with an invisible and undetectable force? We shall instead try to explain all motion that we see—accelerated or uniform—as a characteristic of the RF we occupy.

"Fine! Maybe it will work. Maybe there is no difference between IRFs and RFs. But then why does inertial motion appear straight in one and curved in the other? That's intolerable. If all RFs are to be treated as 'equals,' then there can be no physical distinction between them. The phenomena we observe—inertial motion, in this case—should not allow us to classify RFs into two distinct categories—inertial and noninertial. If these two kinds of RFs are fundamentally different, then the laws of nature would somehow be different for you and for me in the different RFs, and that would violate the first postulate of the theory of relativity.[4]

"No! The laws of physics must be the same in all RFs. If I observe inertial motion to appear curved in my RF, that curvature cannot be something inherent in the motion itself, for if it were, then you could not see a straight path where I see a curved one—that would be impossible. Therefore, the curvature cannot be inherent in the motion itself—it must be inherent in my RF. It is my RF that makes me see the motion as curved, while your RF makes you see the motion as straight. Once we can take into account the peculiarities of our RFs, we can distinguish between the motion itself—the actual physical phenomenon—and its description, which varies from RF to RF.

"The situation isn't all that different from special relativity. There, two observers—one on a train and one on a platform—described two lightning-bolt events differently. Although they differed in their space-time descriptions, they agreed that two bolts had struck. They agreed on the actual phenomenon, but they described it differently in terms of the space and time of their own RFs.

"Because we already know that space and time are different in different RFs, then it should not be too surpris-

---

4. If I say that an object is accelerated and you say that it's moving uniformly, then I require a force to explain the accelerated motion, and you require none to explain the inertial motion. We are then using different laws of physics.

ing that other spatial characteristics differ as well from RF to RF. Because the curvature I see in the motion of a body cannot be inherent in the motion itself, but instead must be characteristic of my RF, then it is the curvature or geometry of space-time that must be different in my RF and yours. I see inertial motion as curved because the space of my RF is curved. You see inertial motion as straight because the space of your RF is straight—or 'flat,' as the mathematicians say. In addition to the relative space-time of the special theory, we must now have relative geometry in the general theory. And thus geometry replaces gravity."

EINSTEIN'S ARGUMENT

Einstein is getting a little ahead of us here. What on earth does he mean by "geometry replaces gravity"?

To follow Einstein admittedly requires patience and diligence. General relativity is far more than an exercise in logic or an elaborate experiment in thought. It represents one of the pinnacles of creative achievement of this century (or any other). Among all the intellectual marvels of twentieth-century physics, many of which we shall explore in this book, Einstein's general theory of relativity stands out for the sheer magnificence of its architecture and the originality and grandeur of its conception. Therefore, it is worth presenting this theory in sufficient detail to expose its aesthetic as well as its intellectual splendors. And because Einstein's argument is long, complex, and subtle, it would be useful to recapitulate its main points.

Einstein began from the principle of equivalence—the idea that you cannot distinguish between an accelerated RF (a reference frame such as a rocket-propelled elevator) and an RF in a gravitational field (an elevator at rest on the earth). He then provided an unambiguous way to characterize the motion of an RF by using the law of inertia. An RF in which the law of inertia is patently obeyed is defined as an inertial RF, or IRF. Next, he noted that uniform inertial motion appears straight in an IRF and curved in an accelerated RF. So the curvature or straightness of *inertial* motion enables us to distinguish between the two kinds of RFs.

But here an ambiguity arises: A gravitational field in an IRF also curves the motion of moving bodies—and curvature, we find, fails to distinguish between acceleration and gravity. If, however, acceleration and gravity are equivalent, and if the curvature test cannot distinguish between them, do we need both? No—we can dispense with gravity, which no one can directly observe anyway. Then we will have no ambiguity about curvature— whether it signifies gravity or acceleration—because we've tossed gravity out. We're still left, however, with a difference between IRFs and garden-variety RFs—the difference between the straight and the curved view of inertial motion.

But is the difference between RFs and IRFs really a physical distinction? Because the natural laws governing the phenomenon we observe cannot be different for different RFs (that's Einstein's first postulate), the curvature we see cannot be intrinsic to the phenomenon. If I use forces and acceleration to describe the motion of an object, and you use only the law of inertia to describe the same motion, then we would be using different laws of physics in different RFs, which is not allowed. So the difference in our descriptions—the curvature of the motion, for example— cannot be associated with the motion itself. Instead, the descriptions must be associated with our RFs—with their motion and therefore with their space-time geometry. (Recall that space-time is a geometrical picture of motion.) One and the same phenomenon appears different in your RF and mine because our RFs have different geometry.

Associating a physical description with an RF also occurs in special relativity, where different space-time descriptions typify different IRFs. The space-time descriptions we use in a particular RF are characteristic of that RF and are not inherent in the physical phenomenon.

Whether I measure a table in inches or centimeters, its length is the same and does not depend on the particular scale of my ruler. Similarly, the different descriptions used in different RFs simply represent different ways of describing one and the same basic phenomenon. If these different descriptions involve the geometry of space-time, then one person's curvature may be another person's gravity. So geometry may replace gravity. Rather than a gravitational field existing *in* space, we may talk of the curvature *of* space. The curvature of motion is not a characteristic of the motion itself but of the curved space in which we observe it.

FLAT AND CURVED SPACES

As in the case of space-time in special relativity, it will be much easier to understand curved space by beginning with two dimensions. The plane surface of a table constitutes a *flat* two-dimensional space; the surface of a basketball or a football represents a *curved* two-dimensional space. We can change a flat space into a curved one by deforming it. If I take a flat piece of paper and roll it into a tube or cylinder, it becomes curved. If I begin with a flat sheet of rubber, I can bend and stretch it into a curved spherical shape or a saddle shape.

What distinguishes a flat surface from a curved one? We creatures who live in three-dimensional space can clearly "see" the difference between the flatness and curvature of a surface. We intuitively distinguish between them. But how can we make our intuition precise and unambiguous? Suppose we want to explain the difference to hypothetical creatures who live in a space of only two dimensions? For instance, one two-dimensional tribe might inhabit a flat plane, and another the surface of a sphere. How could we explain the difference to them? They are confined to two dimensions and are incapable of seeing our three-dimensional space. They cannot "see" or visualize the curvature of the surface in which they live.[5]

We have the same problem seeing and visualizing the fourth dimension—a space one dimension higher than the one we inhabit. If we go back a mere five hundred years to the time of Columbus, we find that many people had trouble conceiving of the earth as a closed curved surface in three-dimensional space. They believed that they lived on a flat earth. They could not observe its curvature and feared falling over the edge if they sailed too far out on the ocean. Only in the twentieth century have we obtained direct visual evidence of our round planet from the pictures provided by astronauts. Only by venturing far into the third dimension of our space have we actually seen the curvature of the earth's two-dimensional surface.

The creatures imprisoned in two dimensions could not see or picture the curvature (or the lack of it) of their surface in three dimensions. How then, could we possibly make them aware of the curvature of their space?

This dilemma poses the question of how to determine the intrinsic properties of the geometry of space. How, for instance, can we detect and measure the curvature of our three-dimensional space when we have no access to the fourth dimension?

The answer lies in our carefully observing the geometrical properties of space and not taking them for granted. More than two thousand years ago, Euclid set down the rules for measuring angles, lengths, and areas on the earth's surface. We call this collection of rules and theorems Euclidean geometry. The very word *geometry* comes from the Greek word for *earth measurement*. Euclid's geometry describes two-dimensional spaces, other than portions of the earth's surface. It applies to any flat plane surface. It tells us, for example, that in any triangle, the

---

5. Once again, for a very intuitive and delightful introduction to the worlds of many dimensions, I refer you to E. A. Abbott, *Flatland* (New York: Dover, 1952).

sum of the angles is always 180°, and that in any circle, the ratio of the length of the circumference to the diameter is always the same number—pi, or 3.14. These facts are demonstrable on any flat plane. But although Euclid's geometry applies to small portions of the earth's surface, it does not apply to any extensive regions or to the surface of the whole globe.

If you draw a triangle on the earth's surface connecting Chicago, Caracas, and Madrid, you will discover that the sum of the angles of that triangle is significantly more than 180°. If you draw a circle on the globe at the latitude of Moscow and measure its circumference and its diameter across the North Pole, you will find that their ratio is considerably less than pi. Any navigator knows and uses these facts every day. Euclid's geometry does not apply to a spherical surface, but somebody else's geometry does—in particular, Georg Friedrich Bernhard Riemann's.

During the nineteenth century, Riemann and other mathematicians formulated the geometry of non-Euclidean spaces—two-dimensional spaces that are not flat or Euclidean but curved. These two-dimensional geometries can be generalized and extended to spaces of three, four, five dimensions, and so on. All of the higher spaces, beyond our own three dimensions, are abstract and imaginary. The rules of non-Euclidean geometry tell us the precise properties of the triangles, spheres, and cubes of these abstract imaginary spaces without ever appealing to visualization or intuition.

In other words, when we use the geometry of Euclid or of Riemann, we are describing the inherent properties of one particular kind of space. If the sum of the angles of a triangle on a two-dimensional surface is 180°, then we know that the surface is a plane even if we can't see it—even if we are two-dimensional creatures confined to the plane. On the other hand, if the sum is greater than 180°, then we know that our surface is something like a sphere (or perhaps an ellipsoid). If the sum is less than 180°, then our surface is curved something like a saddle.[6] We can determine the curvature of any two-dimensional surface (or space of any number of dimensions, for that matter) simply by determining its geometrical rules—and that's something we can do from *within* the space itself. We need not go to some higher space to *see* the curvature. By measuring a triangle or a circle, we can determine the inherent curvature of our space.

In relativity, curvature refers not to a two-dimensional surface but to four-dimensional space-time. We can't even picture the flat space-time of special relativity, so it's hopeless to try to visualize space-time curved in some fifth dimension. We may still use our images of curved surfaces as an analogy, but we cannot visualize curved or flat four-dimensional space-time. We can, however, determine the curvature of space-time by measuring its geometrical properties. Einstein tells us that we observe the geometry of space-time physically as a gravitational field.

GEOMETRY, GRAVITY, INERTIA

We are now in a good position to understand Einstein's conundrum about geometry and gravity. He will tell it.

"So. Where were we? Oh, yes. In observing curved motion we do not distinguish between acceleration and gravity. To make life simpler, then, we dispense with gravity altogether. If we observe any curved motion, we can blame it on the acceleration of our RF and needn't invoke any mysterious invisible gravitational field.

---

6. On the surface of a sphere, the sum of the angles of a triangle is not always the same (it depends on the size of the triangle), but it is always greater than 180°. This is also true for many other surfaces, such as those of a football or the earth flattened at its poles. Such surfaces are said to have positive curvature. If a surface is curved like a saddle or the midsection of an hourglass, then the sum of the angles is always less than 180°, and the surface is described as having negative curvature. The flat plane, all of whose triangles have angular sums of exactly 180°, is the limiting case between surfaces of positive and negative curvature. The flat plane has zero curvature.

"Even the distinction between IRFs and RFs turns out to be a red herring. Inertial motion appears straight in an IRF and curved in a non-IRF. We know that these appearances are not inherent in the motion and are associated with the RFs themselves. Motion appears curved in an RF that has curved geometry. Motion appears straight in an RF that has flat geometry. IRFs constitute a special case of RFs with flat geometry, which is the case we treat in special relativity, not surprisingly. In general relativity, we treat all cases—we consider motion and physics in any kind of RF, which in general has curved geometry. Instead of talking about an inertial RF or a noninertial (accelerated) RF, we talk of an RF with flat or curved geometry.

"The arbitrary distinction between IRFs and RFs disappears. Motion is no less inertial because we happen to observe it from an accelerated RF, or rather from an RF with curved geometry. All motion is indeed inertial or natural. There is no forced or accelerated motion. Natural motion appears uniform (having a constant speed along a straight line) in an IRF with its flat geometry. That same inertial motion appears curved in an accelerated RF with its curved geometry. The law of inertia holds in all RFs, but the appearance of the motion depends on the geometry of the RF in which it is observed.

"Good old Newton concocted a force of gravity to explain departures from the law of inertia. Newton reasoned that Mars, if left to its own devices, would follow a straight path at constant speed. An isolated body slavishly obeys the law of inertia and moves uniformly. Because we never do see Mars obeying the law of inertia, Newton thought that something must cause Mars to move on a curved path. He called that something the force of gravity. But all that Newton or anyone else ever observed was the motion of Mars, never any force of gravity.

"The curvature of Mars's motion is characteristic of our RF, and we need no force of gravity to explain it. Mars is indeed obeying the law of inertia. There is no force bending its path into a curve. The elliptical orbit of Mars is a mere appearance—a geometrical expression of our particular RF.

"The laws of physics are much simpler than Newton imagined. We needn't explain departures from the law of inertia. There are none! All motion is inertial. Objects simply follow the natural contours of space-time. The shape of those contours depends on the RF in which they are viewed. In an IRF, the contours happen to be straight lines. In other RFs, the contours are curved. There is no essential or physical difference between IRFs and other RFs, only a geometrical one. The laws of physics are the same in all RFs, and all motion is inertial."

A NEW LAW OF INERTIA

At this point, you may feel that Einstein's arguments are impressive, but perhaps you're still not quite convinced that we have no further need of a force of gravity to explain the motion of projectiles and planets. After all, you may argue, the motion of Mars can't be just a matter of RF geometry. Is there really an RF in which Mars appears to move uniformly?

The answer is yes and no—and I'm not trying to hedge. Suppose you're in an RF moving along with Mars, i.e., revolving about the sun in the same orbit, just ahead of Mars or behind it. You would observe that in your RF the law of inertia holds, just as astronauts in orbit around the earth observe the uniform motion that characterizes the law of inertia inside their space capsules. Both a falling elevator and an orbiting space capsule are moving freely in the earth's gravitational field.[7] The law of inertia holds in each. In your particular IRF, traveling along with Mars, the red planet would be at rest. Mars certainly obeys the law of inertia on such a local scale.

Now let's suppose you observe Mars on a larger, more

---

7. We will continue to talk about gravitational fields for a while yet. They're just more intuitive than curved space-time.

global, scale. You are in some IRF that is momentarily at "rest" far above the plane of the solar system, in which all of the planets revolve around the sun. When you look down, then of course you see Mars and the rest of the planets moving in elliptical orbits. You are in an IRF that has flat geometry, and yet the path of Mars is curved. Why?

The answer is that the principle of equivalence holds only on a local scale, but not globally. An accelerated rocket cannot be distinguished from an elevator at rest on the earth because on the local scale, in the immediate neighborhood of the elevator, the earth's field is uniform.[8] If you view the earth globally from outer space, however, you would see objects on all sides of the planet falling inward toward the earth's center. The gravitational field of the whole planet does not point in one uniform direction but radially toward the center, like the spokes of a wheel. It is not a uniform field. No rocket can accelerate so as to imitate the *entire* field at once. The equivalence of acceleration and gravity is true only on a small localized scale.

You may think this upsets Einstein's whole argument, because general relativity is based firmly on the foundation of the equivalence principle. It turns out that restricting the equivalence principle to local RFs is no problem at all. Any measurement can be made locally, i.e., in a small, confined RF. For any observation, you can always find, or at least imagine, a small enough local RF in which the principle of equivalence holds. The laws of physics will be the same in all local RFs, and when you want to work on a large scale, you can simply use several small local RFs at widely separated locations around the region of your interest.

Although there is no contradiction between the global and local descriptions of Mars, they are different. We might still ask whether we need gravity to explain a curved orbit on the global scale. The answer is no, but we'll let Einstein tell it.

---

8. See footnote 3.

"The equivalence principle tells us that gravity does not exist on the local scale. The curved motion we observe is a symptom of the curved geometry of our RF. Fine. But on the global scale, we *do* see inertial motion as curved in an IRF. How are we to explain this? I certainly have no intention of resurrecting Newton's gravity, even for such global effects as the orbit of a planet. The force of gravity is no more visible or directly measurable globally than it is locally. It's dead and buried. Requiescat in pace.

"Therefore, the curved orbits of the planets must also be due to the geometry of space-time. Space is curved, not only in an accelerated local RF but also on the global scale. What is responsible for the global curvature? It's the material distributed in space—the stars, planets, comets, dust, gas, even the radiant energy like light, infrared rays, and gamma rays.[9] Matter determines the curvature of space. What we observe as gravitating matter is nothing more than the curved geometry of space. Wherever matter is concentrated and dense—as in the sun—space is curved sharply. Mercury, the planet closest to the sun, has the most tightly curved orbit. On the other hand, wherever matter is diffuse and tenuous, bodies will move nearly uniformly. Pluto, far from the sun's concentrated mass, wings its way in a wide orbit with very slight curvature.

"All the motion we observe is merely a reflection of the curvature of space-time. That curvature is determined locally by the acceleration of our RF and globally by the concentration and distribution of matter and energy in our RF. Moving objects simply follow the contours of the local space-time curvature, obeying a new generalized law of inertia, which applies to all motion without exception."

RUBBER-SHEET GRAVITY

Trying to visualize all of this is not easy. Dealing with space-time is difficult enough. Imagining it curved is all

---

9. Recall from an earlier discussion of special relativity that mass and energy are simply different forms of matter.

but impossible. Replacing gravity by geometry, however, can be visualized—once again with the help of a two-dimensional analogy. We know that gravity does not exist in the flat space of an IRF. In two dimensions, an IRF has the geometry of a flat plane. A pellet sliding on a very smooth, essentially frictionless, plane will glide along on a straight line at a constant speed—a classic case of uniform inertial motion. But suppose this plane is a rubber sheet. (I know it's hard to imagine frictionless rubber, but this is a thought experiment, and we can make the rules to suit our purposes.) Imagine the plane is a horizontal rubber sheet stretched over a large square frame—about ten feet by ten feet. We place a heavy lead ball in the center of the sheet. The ball depresses the sheet, creating a deep well in its center. The sheet now has a kind of curved funnel shape, with the lead ball at its lowest point.

Let's now set a pellet gliding on the frictionless curved surface—not directly toward the lead ball but in a direction parallel to one of the edges of the square frame. The pellet can no longer follow a straight line. There are no more straight lines in the altered geometry of the curved sheet. Instead, the pellet will now circle around in the funnel of the curved surface. If we had allowed friction on our sheet, then the pellet would lose speed and gradually spiral down to the lead ball. But on our idealized frictionless surface, the pellet loses no speed and simply continues to orbit around the central depression.

We shall now imagine viewing the whole process under special conditions. We place ourselves ten or fifteen feet above the stretched rubber sheet. We turn off all the lights, but because we have coated the pellet and the lead ball with luminous paint, we can see them very clearly. We cannot see the surface of the rubber sheet or anything else—just the orbiting pellet and the lead ball. It looks to us like a miniature solar system, with a tiny planet orbiting its diminutive sun in the blackness of space.

We now bring up the lights just a little so that we can barely make out the rubber surface. It is still quite dark. There are no markings on the surface, and because we are directly above the surface we cannot tell that it is curved. Viewed from above, it looks to us just like a flat plane. We see a pellet moving in a curved orbit on a plane surface. "Why doesn't the pellet obey the law of inertia?" we ask. Something is causing the pellet to depart from the expected uniform inertial motion.

If Isaac Newton were in our party, he might well postulate the existence of some invisible abstract force—gravity might be an appropriate name for it—to "explain" the pellet's departure from uniform motion. Since neither Newton nor we can see the true curved geometry of the rubber sheet, we are tempted to look for other explanations. If we allow Einstein to shed some light on the whole scene, we would understand that the orbit of the pellet is curved because of the geometry of the rubber surface and not because of any mysterious force of gravity.

EINSTEIN'S VISION

This thought experiment illustrates the grandeur and sweep of Einstein's remarkable new vision of the cosmos and its natural laws. It represents a radical alteration of the Newtonian "system of the world." Einstein dispensed with Newton's force of gravity altogether, replacing it with the curved geometry of space-time. He generalized Galileo's law of inertia, making it the one and only law governing motion. In the old Newtonian cosmos of absolute space and time, material bodies move uniformly on straight paths at constant speeds, provided they are isolated or far from the influence of other bodies. In response to the invisible force of gravity, objects accelerate and bend their straight paths into curving arcs and orbits. There is one law for "natural," or inertial, motion, and a different law for "forced," or accelerated, motion. The force of gravity, however, is not directly observable: its only manifestation is the accelerated motion it causes. Gravity is pulled out of a hat to explain all the nonuniform motion in the cosmos.

By contrast, the new Einsteinian universe is hewn from four-dimensional space-time, whose curved geometry reflects and signifies the cosmic distribution of matter. The contours of this curved geometry—its valleys and hollows—determine all the possible paths of moving bodies. There is only one law, which describes all motion as natural or inertial. No force of gravity need be invoked to explain curvilinear motion. Einstein's geometry may be complex and subtle, but his physics is simplicity itself—one law of motion and no forces at all.[10]

It took Einstein some ten years to find the correct mathematical expression for his vision. In his quest, he was led to study and master the complexities of non-Euclidean geometry and the mathematics of higher dimensional spaces. It was a long and difficult struggle. Through the many blind alleys, frustrations, and disappointments along the way, he was sustained by his vision and intuition. He never doubted that he was on the right path. In the end, in the throes of his triumphant achievement in 1915, he was rewarded by an ecstatic experience of insight that can only be compared to religious revelation and bliss.[11] Einstein wrote in a letter at the time, "For a few days, I was beside myself with joyous excitement."

MERCURIAL PALPITATIONS

Einstein's delight in 1915 was partly a response to the results of a test he made on his recently completed theory. He applied the new equations to a well-known enigma

10. Einstein was very much aware of the absence of other forces in his theories. Friction, air resistance, magnetism, chemical forces, nuclear forces, and so on can all be reduced to just three elemental forces—the electromagnetic force and two kinds of nuclear forces. Einstein struggled without success through the last thirty years of his life to further generalize his theory to incorporate the other basic forces. This quest for a unified theory of all forces continues to motivate physicists today, and we shall be discussing the new approach later. Although Einstein's general theory of relativity is limited to gravity alone, it applies with amazing accuracy to the large-scale realm of the cosmos, where the force of gravity dominates the motion of stars and galaxies and governs the evolution of the cosmos.

11. See A. Pais, 'Subtle Is the Lord . . .', 253.

involving the orbit of the planet Mercury.[12] One of Newton's crowning achievements was solving the ancient problem of planetary motion. Newton succeeded in demonstrating mathematically that a planet must revolve around the
sun in precisely the orbit that Johannes Kepler had described eighty years earlier in his famous laws of planetary
motion.[13] Beginning from his laws of motion and universal gravitation, Newton calculated that the orbit of a planet
was a perfect ellipse that remained stationary in space.

Newton's calculation actually applied only to the simple case of the sun and a single planet. When you take into
account the effect of all the planets, things are not quite so
simple. In addition to the strong gravitational attraction of
the sun, Mercury is weakly attracted by the other planets.
In fact, Mercury's orbit is neither exactly elliptical nor
perfectly stationary. The orbit precesses, i.e., the entire
elongated ellipse rotates very slowly in space.

You can picture this by imagining a wire in the shape
of an ellipse. A bead moving along the wire represents
Mercury in its orbit. Mercury moves along the wire, completing one revolution every 88 days. But the whole wire
ellipse also rotates exceedingly slowly in space, completing
one full rotation in 225,000 years. This imaginary model
represents how Mercury actually rotates in the Solar System. The rotation, or precession, of the orbit is so slow
compared to the revolution of Mercury that the orbit appears to be stationary, even during the course of many
centuries. But the precession can be detected through very
careful and accurate astronomical observations.

By the middle of the nineteenth century, the precession of Mercury's orbit had been clearly observed by astronomers. It did not quite agree with Newton's theory, even
after all the other planets had been taken into account. The

---

12. For an excellent treatment of the precession of the orbit of Mercury
and its significance in general relativity, see C. M. Will, *Was Einstein
Right?* (New York: Basic Books, 1986), Chapter 5.
13. See T. S. Kuhn, *The Copernican Revolution* (Cambridge: Harvard University Press, 1985), 209–19.

discrepancy was minute, but unexplainable. A rate of pre-cession of 43 seconds of arc per century in Mercury's orbit could not be accounted for.[14] At this rate Mercury's orbit would take three million years to complete one rotation. Faith in the perfect accuracy of Newton's mechanics was so great that this tiny discrepancy had remained a nagging problem for astronomers and physicists for more than fifty years until the advent of general relativity.

In November 1915, Einstein put the finishing touches on his new theory of curved space-time and gravity. To test the theory, he set out to do the same simplified calculation that Newton had done more than two centuries earlier. Einstein calculated the orbit of Mercury, treating it as a single planet in the curved space surrounding the sun. Like Newton, he also found an elliptical orbit. But unlike Newton's orbit, Einstein's was not stationary. It precessed at a rate of exactly 43 seconds of arc per century. This result was a straightforward and inevitable consequence of Ein-stein's new theory. No special assumptions had to be made. The result was so remarkable and thrilling that it gave Einstein "palpitations of the heart."

BENDING THE PATH OF LIGHT

In addition to the precession of Mercury's orbit, Einstein made two other predictions from his new theory. The second prediction concerned the effect of gravity on clocks. In special relativity, we've seen that clocks are affected by motion relative to an observer. In general relativity, motion in an accelerated RF continues to affect the "rate" of time, and because gravity and acceleration are equivalent, clocks are also affected by gravitational fields—or rather, by curved space-time. When a clock is placed in a strong gravitational field (or a highly curved region of space), the clock goes slower than at its normal rate. The best tests of

---

14. An angular degree is subdivided into 60 minutes of arc, which in turn is subdivided into 60 seconds of arc. So 43 seconds of arc is 0.0119 degrees.

this prediction were made in the 1960s and 1970s, and they completely verified Einstein's theory.[15] In one very accurate test, a clock was launched in a rocket, and its rate was electronically monitored and compared to clocks on the earth. The earthbound clocks are in a stronger gravitational field than that of the rocket clock when it is high above the earth. As general relativity predicted, the earthbound clocks were measured to be slower than the rocket clock.

It was, however, the third prediction that catapulted Einstein to fame and initiated his legendary status in the popular imagination. Einstein predicted that a ray of light would follow a curved path in a gravitational field.[16] From an accelerated RF, the straight path of a light ray in space would appear curved (as any straight line in space would appear curved to an accelerated observer). Therefore, the same must be true in a gravitational field, according to the principle of equivalence.

To test this prediction, Einstein selected the strongest gravitational field in our general neighborhood—that of the sun. He calculated that in passing near the surface of the sun, the light from a star on its way to the earth would be bent or deflected by a tiny amount. Naturally, such a light ray would normally be blotted out by the sun's radiance. But during an eclipse, the effect could be observed.

Sir Arthur Stanley Eddington was the most famous English astronomer of his day. Despite the war raging between England and Germany from 1914 to 1918, Eddington became aware of Einstein's work and recognized its revolutionary and profound significance. He prevailed upon the British government to fund a major expedition to observe the eclipse of the sun, which would occur on 29 May 1919. Fortunately, World War I ended before the sched-

---

15. See Will, *Was Einstein Right?*, Chapter 3.
16. For convenience and purposes of visualization in discussing general relativity, scientists often talk about gravitational fields rather than curved space-time.

uled expedition, and Eddington and his colleagues were able to observe the eclipse from two different locations on opposite sides of the Atlantic Ocean. The weather did not fully cooperate with the Eddington expeditions, and fewer photographs were made than had been hoped for. But in the end, Eddington was able to analyze his photographic plates and to determine that Einstein's prediction was correct within the observational errors.[17]

With the announcement of Eddington's results in November 1919, just a year after the end of World War I, Einstein became an overnight sensation. His photograph appeared on the cover of the 14 December 1919 edition of *Berliner Illustrierte Zeitung* (Berlin Illustrated Newspaper—the German *Time* magazine of its day) with the caption, "A great new figure in world history: Albert Einstein, whose investigations signify a complete revision of our concepts of Nature, and are on par with the insights of a Copernicus, a Kepler and a Newton." The *London Times* proclaimed, "Revolution in science . . . New theory of the universe . . . Newtonian ideas overthrown."

Einstein became and remained the greatest scientist of his day. But the story of the full vindication and acceptance of the general theory really occurs much later in the twentieth century, sadly after the death of Einstein in 1955.

MODERN REJUVENATION[18]

The mathematical and conceptual complexity of Einstein's new theory, together with the great difficulty of observing the tiny effects it predicted, prevented general relativity from entering the mainstream of research in modern physics and astronomy for many years. General relativity remained a beautiful but relatively useless idol in the pan-

---

17. For a fuller account of the eclipse adventure, see Will, *Was Einstein Right?*, Chapter 4.
18. The full modern story of the tests and vindication of the general theory of relativity is presented in Will's excellent book *Was Einstein Right?*, already referred to several times. See footnote 12.

theon of physical theories. Beginning in the 1960s, however, a number of factors contributed to a resurgent interest in the theory. New, simpler mathematical techniques had been discovered for applying the theory. Highly accurate clocks and other tools for measuring tiny space-time effects had been invented. Astronomers had discovered new objects in the sky—quasars, pulsars, binary stars—that could not be explained by Newtonian physics. The space program had captured the human imagination and made it possible to perform experiments in the vast regions of the solar system with new miniaturized electronic devices.

Ingenious and ambitious experiments were performed on earth and in the heavens to test the many predictions of general relativity and to explore its use in astronomy and cosmology. The measurement of the deflection of light (the eclipse observation) was greatly improved by using radio waves from distant quasars, which are very remote and enigmatic star-sized objects in the sky that radiate enormous amounts of energy for reasons we don't fully understand. Some of the radiation from quasars consists of waves in the radio portion of the electromagnetic spectrum. The deflection of these waves by the sun can easily be measured as the quasars pass near the sun once a year, as do many stars and constellations in the sky, because of the earth's annual revolution around the sun. Because radio waves are not blotted out by the sun's light, these observations can be made without the pressure of an eclipse observation. Over a period of years during the 1970s, the observations demonstrated excellent agreement with Einstein's predictions.

Perhaps the most accurate test was of a new effect, not originally predicted by Einstein himself but which follows directly from his theory. Because of the curvature of space-time, a radar signal sent over a long distance is "delayed" in time. In other words, a signal traveling on a long path that passes through the curved space near the sun requires a longer flight time than one traveling on an equivalent path in flat space. This time delay of a radar signal became a new critical test of Einstein's general theory and was performed most accurately in 1976–77 during the Viking space missions to Mars.

As part of the Viking program several spacecraft landed on Mars. For most of the world, these landers were doing the remarkable work of observing and photographing the Martian surface and "weather," and they were also making detailed chemical and biological examinations of the Martian soil. For a small group of physicists and engineers, however, the landers on Mars served as pinpoint relay stations for round-trip radar signals sent from the earth to Mars and back. The signals were sent and relayed back to the earth at a time when Mars was on the opposite side of the sun from the earth. Under these conditions, the signals "grazed" the sun in their flight. Thus they passed through a highly curved region of space-time near the sun, which would affect their travel time according to general relativity. Several factors contributed to the accuracy of these measurements: the pinpoint location of the landers on the Martian surface, the careful tracking of the Martian orbit, and a transmission technique to correct for the effect on the signal of the sun's corona.[19] The result was that the Viking measurement of the time delay of a round-trip radar signal to Mars and back verified the time-delay prediction to an accuracy of one part in a thousand—the most accurate direct test of the general theory of relativity ever made.

EINSTEIN'S LEGACY

Einstein's new more general theory surpasses even the old special theory in its remarkable concepts and predictions: Space-time not only stretches and shrinks; it is curved. Gravity is not a force but the effect of the curvature of space-time, so that light and radio waves are "bent," or

---

19. The sun's corona is the outermost part of the solar atmosphere. It contains many ionized particles, which can affect electromagnetic waves that pass through it.

deflected, by gravity. Not only motion but also gravity causes clocks to slow down. And strangest of all is the prediction of black holes, which are the final evolutionary stage of very large stars that have collapsed under their own enormous gravitational attraction. They are called black holes because they form a kind of deep "sinkhole" in space from which no matter, radiation, or light can ever escape.

The general theory of relativity is much more than a revision of Newton's gravitational theory, as important as that is. It is a new and sweeping view of the cosmos based on a geometrical interpretation of natural law. It is the geometry of space-time itself, not any invisible gravitational force, which determines the activity and behavior of planets, stars, and galaxies, and which governs the evolution of the cosmos. Einstein's theory tells us not only how the cosmos acts now but how it has developed in the course of time. General relativity has been the guiding principle for the modern science of cosmology, which for the first time in history has dared to apply the laws and methods of science to the story of the cosmos—to the creation of the whole universe.

The expansion of our universe—the most astonishing cosmological discovery of the twentieth century—was already implicit in general relativity. When Einstein originally applied his equations of general relativity to the whole cosmos, he discovered that the universe could not be static but had to expand or contract. This prediction was too much even for the imagination of Einstein. Like all scientists and thinkers before him, he believed that the universe was unchanging and eternal. Assuming that there was some flaw in his theory, Einstein added to his equations a "correction" term, called the cosmological constant, which made the universe "static." Later, when Edwin Hubble discovered that the universe is indeed expanding, Einstein lamented his introduction of the cosmological constant, which he considered to be the greatest blunder of his life.

The vital connection between relativity and cosmology has remained central to modern physics, and we shall explore it later. General relativity has also greatly influenced the development of our latest ideas in the microworld—the realm of atoms, protons, and electrons. The evolution of the universe from its big bang explosion inevitably involves as well the evolution of matter. To understand how the big bang launched the ongoing expansion of the universe, we must also understand how matter evolved from its primal forms—the elementary particles and their antecedents. General relativity is inextricably linked not only to cosmology but also to the physics of elementary particles.

In the twentieth century there seems to be a suggestion of a remarkable merger of two formerly separate and distinct realms of physics—the macrorealm of cosmology and the microrealm of elementary particles. What at first sight promises to be a successful marriage may in fact turn into a shotgun wedding! Elementary particles obey the laws of quantum theory, while cosmology obeys the laws of general relativity, and a greater antagonism between two theories is hard to imagine. The modern quest to unify all of physics from the macrorealm to the microrealm has so far been frustrated by a deep incompatibility between the two giants of twentieth-century physics—general relativity and quantum theory.

Is this conflict merely another puzzle to be solved? Will a new theory—a new paradigm—ultimately transcend both giants and make them obsolete, as has happened to theories before them? Or does this discord represent such hostility and opposition between the camps of physics that science cannot effect a resolution and still survive intact?

Stay tuned for future episodes in the war of the worlds.

CHAPTER 4

Metaphors of Space and Time, Part II

THE SPECIAL THEORY OF RELATIVITY presents a new space-time metaphor that reinterprets the motion of objects and the connection between causes and effects and portrays them in geometrical terms. The finite value of the speed of light has forever altered the old Newtonian notion of instantaneous cause and effect. Science never deals with the metaphysical questions of *why* any kind of motion in space or transmission of causal information is possible in the first place. In effect, the scientist accepts space-time, along with matter and energy, as the *givens* of the physical world, and their ultimate nature remains a mystery. By tradition, science eschews metaphysical questions and treats them like the plague. But the theories and models of science have unavoidable existential, metaphorical, and even spiritual implications, which become very apparent and insistent in general relativity, as we shall see.

SPACE-TIME CAUSALITY

Newtonian cause and effect have been altered by relativity theory because of the special role played by the speed of light. No matter or energy can travel faster than light, and therefore no information can travel between two points in the material world any faster than light. In the old Newtonian framework, the force of gravity was presumed to travel instantaneously from one body to another: If the sun should suddenly cease to exist, the earth and all the planets would respond instantaneously. They would immediately begin to obey the law of inertia. Each planet would instantly move off on a tangent to its elliptical orbit and travel along a straight line at a constant speed. At least, that would be the Newtonian scenario.

In relativity, however, any change or alteration in a gravitational field cannot be transmitted instantaneously from place to place but travels instead at the speed of light. Light itself is an electromagnetic wave, i.e., a traveling disturbance or variation in an electromagnetic field. Similarly, any change in the pattern of a gravitational field also moves at the speed of light, like an electromagnetic wave. The vanishing gravitational field that would be caused by the sudden hypothetical demise of the sun would propagate out from the sun's former location at the speed of light. To cross the ninety-three million miles between the earth and the sun would take a little more than eight minutes. The earth and all its inhabitants would continue to follow their elliptical orbit for eight minutes after the sun's disappearance. For those eight minutes, we would all be mercifully ignorant of our terrible impending fate. The last rays of the sun's light would reach us during those final eight minutes before the descent of eternal night.[1]

Of course, certain causes—an explosion, for example—may produce a distant effect through the displacement of material objects, such as flying debris hitting a wall. Then, since matter always moves at a finite speed, there is clearly a time delay between cause and effect. But

1. For a fuller account of relativity and causality, see R. S. Jones, *Physics as Metaphor* (Minneapolis: University of Minnesota Press, 1990), 100–113.

while Newtonian physics allows for instantaneous cause-effect relationships, relativity rules them out. So causality is modified in relativity, but not in any essential way. The cause always precedes the effect,[2] and there is an inevitable time delay between them in all cases.

The essential idea of causality is that of an initiating event transferring matter or energy to a different location where the matter or energy induces a second event to occur. I dial your number. Your phone rings and you pick it up. I initiate something to which you, some distance away, respond. Another example—a part of the ocean floor sinks, setting up a tidal wave that sweeps across the Pacific and inundates the Peruvian coast. Cause and effect.

Causality is fundamentally based on the concept of motion. An electromagnetic wave moves through space or through a telephone line and causes some effect at a distance. Pitching a ball toward home plate, making ripples that spread on a lake, or sending radio signals from ship to shore all illustrate the movement of matter or energy that is involved in causal connections. Two events that occur simultaneously at a distance from each other (in some IRF) cannot be related causally because no energy or matter can move between them instantaneously (i.e., at an infinite speed). A causal connection between two events implies the motion of matter or energy between them. There can be no causality without motion.[3]

There can be no motion without space and time. Motion means a change of spatial position in the course of time—distance traveled in elapsed time; an interval of space and a corresponding interval of time. The very notion of causality hinges on the nature of space and time. Because space and time are inextricably linked in relativ-

2. In relativity two events that are causally related will never have their time order reversed in any RF. The time order of cause and effect is always preserved in relativity.

3. Even causal events that are very close to each other require some interval of time, no matter how small, for information to get from one to the other.

ity, causality is part and parcel of the structure of space-time. A Buddhist monk in a deep meditative state that transcends space and time can have no experience of causality. There can be no causality without space-time. Causality, which is at the heart of all physical or natural law, is a kind of artifact or construct of space-time.

METAPHYSICAL SPECULATIONS

What is the ultimate nature of space-time? Questions such as this, which deal with metaphysical speculations about the essence of things, are not considered a legitimate part of today's scientific inquiry. Science generally eschews metaphysics and limits itself to matters that are empirical, measurable, and observable. And yet, we cannot refrain from asking such questions. In Chapter 2, we examined the representational and constructed nature of space-time, and we also questioned the reality of the three-dimensional space we think of as filling our physical universe. We considered space as a metaphor of relationship—a web of connections, distances, orientation, separation, identity—and even the foundation or platform of existence itself.

All civilizations and cultures have had their favorite metaphors for space and time. The current candidate of modern western European science and culture is a kind of vast, sterile void—a space characterized by the absence of everything and yet which permeates the universe like some odorless, invisible perfume. But "permeates" is misleading. You can only permeate or fill a space—a volume—that is already there. Space is more like the scaffolding, the lattice, the web that supports nature. How can it be a scaffold if it is immaterial and if it has no rigidity? In modern cosmology, it even expands. But what is there to expand? How can emptiness stretch?

It's no wonder that scientists have discarded such questions from science and relegated them to the realms of metaphysics and philosophy. These are not matters on which the methods of science can shed any light. In thousands of years of speculation, philosophers, mystics, and sages have made no progress with them. We know no more about the essential nature of space and time than did the Greeks. Of course, science does deal with space and time on its own terms. We can describe gravitational fields and moving bodies in space, we can send astronauts and radio signals through space, and we now understand the relativity of space. Surely, this must be progress. But how easily we are fooled by language into thinking we are saying something. We are especially fooled by those wonderful little prepositions—in, through, of. What does it really mean to talk about something in space? What does *in* mean? What is there in the physical world that is not in space? What is "inness" apart from the notion of space? Even when we use the word *in* abstractly—in Shakespeare, in 1905, in absentia—there is still the idea of containment, insideness, withinness. But we are playing games—merely substituting one space metaphor for another.

Prepositions are the words in most languages that express spatial relationships—in the house, out of the country, through the tunnel, over the rainbow. But they are words, signs, symbols—no more. They do not explain space or help us fathom its mystery any more than science does. When we use prepositions to signify spatial relationships, we simply describe experience that we take for granted. *In* would mean nothing to a disembodied space-transcendent being, any more than *red* means anything to a blind man.

To say that a gravitational field is in space is to say nothing. Where else could it be? A gravitational field is itself a condition or property of space. Of course it's in space. To say, as scientists do, that we describe gravitational fields as a "function of" spatial coordinates means that we can write a mathematical expression for the strength and direction of the force of gravity in terms of symbolic quantities that represent positions in space. At such and such a position, the force is so big and points this way. At some other place, the force is this big and points that way. So we describe properties of space in terms of space.

"Hold on," Newton might well say in protest. "A force is not a property of space. It is something we can measure." Newton hesitates for a moment before continuing more reflectively. "Of course, we cannot see a force itself or measure it directly. We do not see the earth attracting a falling rock. We see only the rock's fall. We can measure only the effect of a force in terms of the motion or change it causes in some material body—the fall of a rock or the stretch of a spring." And then, after another pause, "But to measure these things, we must of course describe the falling motion of the rock or the displacement of the spring. So we are measuring . . . what else—why, space, of course. Ah . . . perhaps I spoke too soon. It does seem that the measurement of a force is equivalent to the measurement of space."

Perhaps we all speak too soon and too glibly about space and time. They represent deep mysteries—metaphors for the very nature of being and existence. All the laws of science do not even begin to explain the mystery. They do, however, give us some insight and an important viewpoint, and they are certainly valuable.

THE CURVED SPACE-TIME METAPHOR

In the general theory of relativity, Einstein dispenses with the law of gravity and explains the motion of bodies in terms of the curvature of space-time. A freely moving body follows the natural contours of space-time, and so its motion may appear curved in one RF and uniform in another. The curvature of space in turn is determined by the configuration and concentration of matter and energy. Instead of talking about the gravitational field in the vicinity of the sun, we now speak of the curvature of space in the sun's neighborhood. The large concentrated mass of the sun "produces," "is responsible for," "is equivalent to," "is manifested as" the curved geometry of the space around it. It's difficult to know just how to put into language the mathematical expression that equates the distribution of matter with spatial curvature. If the sun's mass is equivalent to the curvature of space, then what form does the sun assume in space? Is it there as matter or as a region of high curvature? Do the equations of general relativity dematerialize matter? Do they geometrize it? It's one thing to replace gravity with geometry, but are we replacing the sun with geometry?

The problem of representing the sun itself (and not just its gravitational field) was the roadblock that stymied Einstein in his efforts to further generalize his theory. Gravity operates on the large scale of the universe—in astronomy and cosmology—and even down to the scale of the earth and the objects on it. Matter is not held together, however, by the force of gravity. As we move into the realm of everyday objects and the elementary particles that make them up, gravity is far too weak to play a significant role. The molecule, the atom, the nucleus, and the quark are dominated and governed by the electromagnetic and nuclear forces. Einstein's thirty-year struggle to incorporate just one of these forces—electromagnetism—into an enlarged geometrical theory like general relativity was entirely unsuccessful.[4] The sun and, in fact, all matter, are also subject to the nongravitational forces—the electromagnetic and nuclear forces. The sun cannot simply be equated with its gravitational field. Because matter cannot be explained in terms of gravity alone, and because general relativity deals only with the force of gravity, then relativity theory can never give a complete explanation of the sun or any real material body. Relativity alone cannot account completely for the entire physical existence of the sun because in relativity it is only gravity, and not electromagnetism and nuclear cohesion, that is equivalent to geometry. The sun is more than the geometry of its gravitational field, more than simply a curved region of space-time.

---

4. The quest for a unified field theory of all the forces of nature continues today, but using a different approach from Einstein's (see Chapter 18). Combining gravity with the other forces remains a very tough nut to crack.

THE MATTER OF SPACE

In contemplating general relativity one can still raise tantalizing speculations about the materialization of space. In a hypothetical universe governed only by gravity, would matter be mere geometry? When we speak of the geometry of space, do we mean something separate and distinct from space itself? Do my form and shape exist apart from me? Does space exist apart from its curved geometry? If matter (at least gravitational matter) is equivalent to the geometry of space, is it then equivalent to space itself?

If nothing else, general relativity reverses the trend we spoke of in Chapter 2. Science has gradually scoured from space any traces of material substance, organic structure, mental activity, or meaning. Newtonian space is a sheer empty void. With Einstein, space once again seems to be replete with substance and structure. Although we can hardly fathom what it all means, we no longer have gravitational fields inhabiting otherwise empty space. Instead, we have matter and gravity incorporated into the structure—the very fabric—of space. We cannot see and touch geometry, but perhaps we can feel some kind of security, some vague familiarity, in this abstract, yet material, space of Einstein's.

In fact, the fantastic alloy of space, time, and matter that we encounter in general relativity is hardly a new metaphor. Ancient, primal, and non-Western cultures developed similar notions long ago. There was, for example, the organic cosmos of Hellenic Greece in which all aspects of existence on earth and in the heavens were connected and interrelated in a great organic hierarchy of Platonic forms and the four elements—earth, water, air, and fire. The cosmos was filled with intelligence: it was a plenum of mind and wisdom.[5] It was a realm of support and nourishment, not an empty void without form, shape, or succor.

---

5. See O. Barfield, *Saving the Appearances* (London: Faber & Faber, 1957), Chapter 15.

There was the *causal-body* of Hindu philosophy.[6] This is an amalgam, not only of space and time but of all thought and feeling and of all potential possibilities. Embedded and recorded in the matrix of the causal-body is all of history and all possible futures and histories and all the thoughts, ideas, and knowledge that are known or can ever be known or might ever be known. This conception of space-time is like a great hall of records, a fabulous library whose endless corridors house all knowledge, being, and meaning. A far cry from the echoless empty caverns of Newtonian space.

Bare empty space is a relatively new conception—one that would be alien, unfamiliar, and incomprehensible to people of primal and ancient cultures. If we think of these earlier views of space and time as naive, misinformed, and anthropomorphic, we must at least recognize the sense of home, security, and support the earlier conceptions of space gave to those who believed and dwelt in them. We have given up this nourishing quality of space—of our abode—at our peril. We pride ourselves on our no-nonsense attitude, on having risen above superstition and wishful thinking, on having no need for illusory security, on facing the world and existence realistically and soberly. But the instabilities and deep flaws of today's world bespeak a false sense of confidence and a counterfeit optimism.

Both relativity and quantum theory (as we shall see) have begun to replenish space with fullness and meaning and have begun to combine space, time, and matter into some new, strange—but enticing—alchemical brew. Is there some mythic or spiritual meaning hidden in these metaphorical amalgams? Can general relativity shed any light on the matter?

---

6. See A. Daniélou, *The Gods of India* (New York: Inner Traditions International, 1985), 242.

CHAPTER 5

The Mechanical Worldview

IF RELATIVITY AND QUANTUM THEORY are revolutions, then what did they revolt against? During the eighteenth and nineteenth centuries, the prevailing view in physics (and in science generally) was mechanistic—the universe was pictured to be a huge machine, operating with clockwork precision. Even the celestial spheres of the ancient Greeks, which kept the planets revolving in their orbits, had been imagined by medieval thinkers to have mechanical gears and to strive against friction. The epitome of the mechanical worldview, however, was Newton's "system of the world."

DESCARTES AND GALILEO

Many were the giants upon whose shoulders Newton stood.[1] But his immediate intellectual predecessors were Galileo and Descartes, who set the stage for Newton's cos-

---

1. "If I have seen further [than you and Descartes] it is by standing upon the shoulders of Giants." Letter from Newton to Robert Hooke (5 February 1675/1676?).

mic drama. Galileo said that the book of nature is written in mathematical language. Mechanics had long been the study of the natural laws of moving bodies, but Galileo insisted that the basic concepts of mechanics must be mathematical.[2] This in turn required that only quantitative, objective characteristics of things—what Galileo referred to as "primary qualities"—could be considered in the science of mechanics. The number of objects, their size and shape, and their position and state of motion in space could all be quantified and made a part of natural law. Such "secondary qualities" as redness, sweetness, noisiness, and foul odor, however, depend upon the senses and "reside only in consciousness."[3]

Galileo thus studied and described motion in precise mathematical terms. He discovered and stated the law of falling bodies—that their speed was proportional to the elapsed time and did not depend on their weight. This was truly a mathematical analysis of motion.

At the same time, Descartes reinforced Galileo's ideas by equating the knowledge of nature with the knowledge of mathematics. Descartes believed that physics was the science of moving forms of space, just as geometry was the science of resting forms of space. He, too, insisted that objective nature consisted only of the mathematical aspects of objects—size, shape, and quantity. All animation, internal spontaneity, and purpose were denied. *Mechanics* meant capable of being imitated in a mechanical model. There was no difference between a running clock and a growing tree.[4]

Descartes also formulated the first clear statement of the law of inertia: The natural motion of an isolated body is uniform (i.e., along a straight line at constant speed—as we discussed earlier). Departures from uniform inertial

2. See E. J. Dijksterhuis, *The Mechanization of the World Picture* (Princeton: Princeton University Press, 1986), 499.
3. See Galileo, *Discoveries and Opinions*, ed. S. Drake (Garden City, N.Y.: Anchor, 1957), 274.
4. Dijksterhuis, *The Mechanization*, 414–15.

motion were attributed to the pushes or pulls exerted by other bodies. Descartes never succeeded in developing a full mathematical science of mechanics, but his ideas were to become the starting point for Newton, whose new mechanics of bodies accelerated by forces was essentially a grand mathematical design to rationalize all motion that departs from the law of inertia. Newton stunningly fulfilled the promise of Galileo and Descartes by constructing a completely mathematical theory of planetary and terrestrial motion. It is impossible to exaggerate the effect and influence of Newton's seminal accomplishment. Ultimately, it swept through the world and opened a new era in our understanding of nature as well as our power over it.

MECHANICS AND SEMANTICS

The new mechanics of Galileo, Descartes, and Newton had come a long way since the time of ancient Greece. The contrast couldn't be more extreme. Greek motion was always considered to have a purpose: it was due to "final" causes, as Aristotle would say.[5] In other words, the motion was governed by its final aim or purpose. This attribution of purpose or design to nature—teleology—was typical of Greek philosophy. Despite our distaste for teleology and its rejection by modern science, it was precisely the intuitively intentional character of Greek motion that made it comprehensible, meaningful, and natural to the ancient mind. Motion was understandable in one of two ways: Either a body would seek out a sympathetic position among like substances somewhere within the natural vertical hierarchy of the four elements—earth, water, air, and fire—or else it would move as the result of an urge or desire on the part of a "mover" to perform an act or to bring about a change. A rock falls naturally downward toward its own earthy realm at the center of the cosmos, or it is propelled

---

5. See, for example, G. F. Parker, *A Short Account of Greek Philosophy* (New York: Harper & Row, 1967), 143.

in some other direction by an outside mover. Greek motion, as has been mentioned, always had some purpose behind it—it restored an imbalance or fulfilled a desire. It had an animating rationale or drive that made it seem intuitively meaningful and therefore understandable.

Newtonian motion, on the other hand, is neither intuitive nor purposeful. It is caused by some agent or principle—a force, for example. This is called an "efficient cause" as opposed to a "final cause" in Aristotle's terminology.[6] Natural uniform motion has no inherent logic or rationale behind it. It doesn't fulfill any cosmic scheme; it simply is. The law of inertia, despite its mathematical precision and predictive power, simply states without explanation what the natural state of affairs is in the universe. Isolated bodies simply move forever *of their own accord* along straight lines at constant speed. From the lay point of view, this can seem quite arbitrary, as if proclaimed by fiat. Like many laws of physics in the modern era, the law of inertia is counterintuitive and makes no sense (no common sense, that is).

Someone might object, saying that intuition is simply wrong and must be discounted and that the accuracy and power of science fully justifies its methods. But we must make a distinction here between efficacy and meaning. In the first place, which criteria are most appropriate in making a critical assessment of science? Clearly, Newtonian mechanics (and its twentieth-century revisions) is *empirically* superior to all earlier treatments of motion. The remarkable quantitative accuracy of Newtonian mechanics (not to mention its heirs—relativity and quantum theory) has made physics and its technological derivatives indisputably preeminent in history and in contemporary civilization.

---

6. An efficient cause is the direct, immediate, external agent that produces a change. If I push you, my hand is the efficient cause regardless of what my intention (final cause) is. See also, for example, Parker, *A Short Account*, 143.

But are predictive accuracy and control over nature the only legitimate criteria by which to evaluate the influence and significance of science? This is a fundamental question, especially if we want to view science from a larger cultural perspective—from a more humanistic perspective. Our concern is not simply with the efficacy of science but with its ability to help human beings rationalize and make sense of their lives and experience. The philosophies, myths, and cosmologies of earlier times may seem completely inadequate and naive nowadays because they fail to provide an accurate and verifiable description of the natural world. But based on their ability to add meaning and significance to human life, they may indeed score higher than does modern science.

The fact that Newton's laws are stated in precise mathematical form makes them no more reasonable or compelling from an aesthetic, instinctive, or philosophical point of view. Mathematical laws provide us with great predictive powers, but not with intuitive comprehension or meaning. To a mind that hungers for innate meaning or intuitive understanding, the mathematical laws of motion offer precious little nourishment. However misguided Greek ideas about motion may be, they did evoke a satisfying sense of meaning that seemed compatible with the human sense of the cosmos. This uniquely Greek character is almost entirely absent from Newtonian and modern mechanics.

NEWTON'S SYNTHESIS

In 1987, we celebrated the three hundredth anniversary of the publication of *Principia Mathematica Philosophiae Naturalis* by Isaac Newton. This pivotal book is generally recognized as the greatest scientific treatise ever published. Newton presents nothing less than a new "system of the world." We are given the simple laws of motion that apply equally well to falling apples and revolving moons. These laws are postulated to explain the motion of all material bodies. By way of proof, Newton uses the power of mathematics to unravel, once and for all, the ancient mystery of the motion of the heavenly bodies.

The departures of the planets from the regular steady motions of the stars had puzzled observers of the skies since before the time of the Greeks. By Newton's time, the seemingly erratic motion of the planets had been consolidated and summarized by Johannes Kepler in the form of three simple mathematical laws. Essentially, Kepler's laws state that the planets follow elliptical paths about the sun according to mathematical rules that relate their velocity, their distance from the sun, and the period of their revolutions about the sun. Newton went beyond the mere description of planetary motion to explain *why* the planets move as they do. If the planets had existed in empty space, far from other heavenly bodies, they would move uniformly, inexorably obeying the law of inertia. But in the solar system, the planets feel the effect of the massive sun that tugs on them with Newton's postulated *force of gravity* and bends their straight-line trajectories into elliptical orbits.[7]

In Newton's scheme, the action of a force is responsible for departures from uniform inertial motion, whether it's the accelerated fall of an apple or the curved path of the moon. It may seem quite mystifying that the earth should "pull" on objects, speeding them up and bending their paths. But Newton's analysis was very convincing. He stated his principles in the language of mathematics, and then he went on to use the grammar of this language to describe precisely the motion of a falling apple or of the revolving moon and planets—in fact, of any material body. Newton's generalized rule states that *forces cause departures from uniform inertial motion.*

But how forces arise and perform their functions were questions he did not answer.

---

7. Clearly we are in a period here long before the curved space of general relativity.

CAUSA NOSTRA

The word *cause* in this context provides the fundamental model for the notion of causality that has been used in science for the past three centuries. At the same time, this usage displaces and disenfranchises the older, more traditional idea of a cause as a psychological reason or intuitive motivation behind behavior. *Newtonian* (and therefore scientific) *causation* means that some agent, such as the force of gravity, bears a mathematical relationship to some effect, such as the acceleration of a planet. When we say that a force causes acceleration, we have in mind a mathematical equation that relates a change in velocity (acceleration) to a given force. The force of gravity on a falling body, for instance, is mathematically equal to the acceleration of the body (multiplied by its mass).

All of this works remarkably well and has been verified in countless observations, but does it actually explain anything about the nature of force, acceleration, or motion? Does Newtonian causation explain the causes of things as the Greeks understood them, or, for that matter, as today's average person on the street understands them?

Suppose, for example, I say that an infectious disease is caused by some invasive microorganism. I mean that this microbe is harmful to the human body and that when it invades human tissue, it will cause a painful or inflammatory reaction. It is not any knowledge of biochemistry here that aids my understanding. Rather, the intuitive notion of a foreign substance that harms the human body makes a certain amount of sense and satisfies the desire for understanding.

I do not deny the value and efficacy of a medical or biochemical explanation. What I do say is that a chemical explanation does not make it *intuitively* clear to most people *why* a certain toxin or organism is harmful to the body. A biochemist might object and say that intuition is not the issue here: research has scientifically demonstrated the cause of the disease and only a trained biochemist can fully understand the explanation. In examining the deeper meaning of causality, we cannot, however, simply reject intuition as contrary or irrelevant to science. We need to view the conflict between scientific efficacy and intuitive meaning from a larger, more philosophical and humanistic perspective than that of science itself.

When we say that a force causes acceleration or a change in uniform motion, we give neither an explanation nor do we reveal an ultimate cause. How does force change speed and direction in the process of accelerating a body? What in the nature of force enables it to affect motion? And ultimately, what is a force?[8] The answers to these and other related questions are never given by Newton's "explanation" of motion. In fact, the only logically valid definition of force in Newtonian mechanics is *that which causes acceleration*. This well-known tautology in modern science is commonly ignored and dismissed with the observation (more aptly, the quip) that physics works and logic be damned. The layperson (to say nothing of the fledgling science student) is often intimidated and left with a troubling sense of dissatisfaction and embarrassed inadequacy for failing to *understand* such a pragmatic scientific explanation.

This point is very difficult to argue nowadays when most explanations are cast in scientific form, and other methods of knowing and understanding (e.g., psychological, intuitive, meditative) are often rejected or even ridiculed as superstitious, subjective, and anthropomorphic, i.e., modeled inappropriately on human motives and behavior. The uneasiness that laypersons may experience in the face of so-called scientific explanation is frequently attributed to ignorance or the lack of a "scientific mind." People may suppress or conceal their uneasiness and thus fail to place some blame on the explanation itself for not

8. The inability to give an adequate answer to this question was part of Einstein's argument in general relativity for rejecting the force of gravity.

satisfying the intuition. It is not simply a matter of subjectivity and personal idiosyncrasy that scientific explanation is so counterintuitive.

ENGINES AND DASHBOARDS

Suppose, for example, that you ask me how a car works. I respond by telling you how to turn on the ignition, step on the gas, and steer with the wheel. You might object that I am explaining how to drive a car and not how it works. I counter that driving a car is what you really need to know and that the question of how a car works is largely irrelevant. I could reinforce this view by demonstrating that turning on the ignition, stepping on the gas, and steering with the wheel is precisely how one drives the car. After a little thought (and perhaps some driving experience), you'd have every right to complain that this explanation is of no value to you when the car breaks down. You would realize that knowledge of how to drive a car—"dashboard knowledge"—is no substitute for "engine knowledge."

Just as the prescription for driving a car is quite different from an understanding of how a car works, so the description of motion, however precise, is entirely different from an understanding of the ultimate causes of motion.

The Newtonian notion of a force acting at a distance is altogether different from the Greek notion of causality, which suggested a kind of intelligence and communication among the organic parts of the living Greek cosmos. Instead of a natural kinship and affinity among things, Newton introduces a mechanistic scheme to link inanimate unrelated objects and events. In modern science, causality typically links events by describing how information is communicated between them in some material form (electromagnetic energy, for example). Cause-and-effect relationships amount to abstract, mechanistic connections among things, as opposed to the metaphorical, organic kinships that prevailed in the Greek cosmos.

Because Newton had given a rational (mathematical) explanation for the notorious and ancient problem of planetary motion and went on to provide unheard of and unanticipated predictive power and control over nature, his system of the world completely won us over. We lost sight of the original question, asked by Greek and other earlier civilizations, viz., what causes motion and change.

Newton himself realized that his explanation of planetary motion was far from complete. Nothing in his treatment tells us how the mysterious force of gravity arises from the matter of the sun, or how it is transmitted across millions of miles of empty space, or how it "grabs" the planets and deflects them into elliptical orbits. Despite these obvious inadequacies, Newton's scheme does *mathematically derive* the planetary orbits from an ad hoc force of gravity. Because of this, Newton and the rest of us have fallen, just like the planets, under the alluring attractions of the powerful force of gravity.

FORCES OF MODERNISM

Before continuing our discussion of Newton's mechanistic system in the eighteenth and nineteenth centuries, we will look ahead briefly to see how the thorny notion of force is handled in the twentieth century and whether modern notions of causality are any more intuitive. In the twentieth century, as we have seen, Einstein would reevaluate the need for Newton's arbitrary and unobservable force of gravity and finally discard it. In his general theory of relativity, Einstein formally proposes that there is nothing unnatural about accelerated or nonuniform motion. He attributes departures from uniform motion to the geometrical properties of space, rather than to some trumped-up force. Moving bodies follow straight paths in flat space and curved paths in warped space. It is the geometry *of* space, not forces *in* space, that determines the motion of bodies. The distinction between uniform and accelerated motion disappears, and with it goes the need to introduce the concept of a force of gravity to explain unnatural motion.

All motion is natural or inertial. Moving bodies simply follow the contours of space, which in general are curved paths.

It may seem a moot point to transfer the causes of acceleration from an arbitrary definition of force to one of space. In Einstein's rejection of the notion of force and the reclassification of all motion as natural, however, we cannot fail to recognize a deep aesthetic gesture towards symmetry and simplicity and away from arbitrariness. Einstein's "explanation" of motion may, like Newton's, still seem intuitively wanting, but it is more "Greek" and satisfying than Newton's in that it establishes a kind of kinship or affinity between all forms of motion and the inherent or elemental character of space. In a limited sense, Einstein goes the Greeks one better (not to mention Galileo) by eliminating the idea of unnatural motion altogether.

The force problem is handled in quite the opposite way in quantum theory. (We shall explore quantum theory in detail in later chapters, but here we will anticipate the quantum treatment of force to see how well it fares as a causal explanation.) Newton had placed forces in a kind of middle ground somewhere between space and matter. A field is clearly neither purely material nor purely spatial. Einstein etherealized force by pushing it completely into the background—into space itself. But in quantum theory, force is effectively materialized.

In quantum theory, the mystery of how forces arise from matter and are transmitted through empty space is explained by the emission and absorption of particles, called *field quanta*, which are like messengers that carry or transmit the effect of the field. An electron, for example, revolves around a proton in the hydrogen atom. Instead of saying that the electric field of a proton reaches across space to attract and deflect an electron, quantum theory says that electrons and protons attract each other by exchanging photons, which are the field quanta of the electromagnetic field. There is no mysterious field in space, only material particles—photons—traveling back and forth between the attracting particles. The electron is made to orbit the proton, not by some ethereal force in space but by its recoil as it emits and absorbs photons. So the motion of two interacting particles—an electron around a proton or a planet around the sun—occurs because of a series of recoils as field quanta are exchanged.[9]

There are also field quanta to explain the other forces of nature, as well—the nuclear forces and presumably gravity.[10] These quanta are still something of a mystery to the mathematically uninitiated, but they are certainly a form of matter and energy. They travel through space like other particles or waves, and this seems to solve the problem of Newton's instantaneous action at a distance.

As an intuitively satisfying explanation, field quanta are not much of an improvement over Newton's forces. They may even be more problematic since the erratic and unpredictable creation and destruction of these quanta are attributed, with no further explanation, to what is called the basic random quantum activity of matter. In other words, quantum theory assumes, without explanation, that quanta themselves are capable of spontaneous generation and movement. To explain the motion of matter, quantum theory assumes the motion of matter! This is a logical fallacy. To define a cat in terms of cat tells us nothing. Despite the enormous pragmatic successes of quantum theory, its logical and philosophical problems have turned out to be far more subtle and perplexing than those of Newton's mechanics ever were.

Ultimately, relativity and quantum theory are just as mechanical as Newtonian theory, though considerably more abstract. There is a far wider and deeper gap between Aristotle and Newton than between Newton and the twentieth-century physicists.[11] Modern ideas of motion, causal-

---

9. We shall explore this mechanism further in Chapter 18.
10. The graviton, or field quantum of the gravitational field, if it exists, will be exceedingly difficult to detect.
11. See Dijksterhuis, *The Mechanization of the World Picture*, part V, section 9.

ity, and mechanics follow in the footsteps of the Newtonian counterintuitive tradition, and, in fact, carry it to a further extreme of abstraction.

Incidentally, the difference in approach between relativity and quantum theory on the matter of force is typical of the conflict between these largely incompatible systems of thought. One of the great struggles of twentieth-century physics (vigorously renewed in the 1980s) has been to reconcile these two theories, largely by attempting to recast Einstein's geometrical theory of gravity as a materialized quantum field theory. In relativity, the force of gravity is spatialized, while in quantum theory, the other basic forces—electromagnetism and the two nuclear forces—have been materialized and are described in terms of material particles (field quanta) traveling in space.

So far it has proven extremely difficult to apply the quantum method to Einstein's geometrical gravity. Relativistic gravity is not a field in space but space itself, which cannot simply be replaced by particles. How, for example, can particles traveling through space be the same as the space they are traveling through? (Once again such elusive prepositions as *in* and *through* conceal a wealth of complex assumptions about space.) Proponents of the new superstring theory claim possibly to have a "shotgun wedding" between relativity and quantum theory in their sights. If this is so, it may lead to a unified description of the knotty relationship between matter and space and the concomitant role of force. But will it give us any deeper understanding of the nature of these things?

Saving the Appearances

Newton's theory of planetary motion is the veritable prototype for all mechanical models. It is the apotheosis of the mechanistic view of the world—the fulfillment of Descartes' dream. The whole cosmos operates like a machine—like a clock! Medieval thinkers, following Aristotle, had attempted to explain the rotation of the crystalline celestial spheres in mechanical terms.[12] Now Newton had provided a metaphorical cosmic clock that, once wound up and set into motion (presumably by God), would run like a perpetual-motion machine with perfect but mindless precision. As scientists and mathematicians began to study and apply Newton's system, however, they began to discount and reject the metaphorical character of the Copernican model of the Solar System and instead began to place more and more emphasis on Newton's mathematics as an actual description of reality. This subtle but profound shift in viewpoint amounted to a new role for mathematics, in fact a new relationship between mathematics and the natural world.[13]

For the Greeks, geometry had served to "save the appearances," i.e., to reconcile the changing and impermanent character of the sublunary world of appearances with the perfect order, harmony, and permanence of the ideal celestial realm. Plato believed the ideal realm to be the reality behind the illusion of the everyday world. So for Plato, and the later Neoplatonists, mathematics constituted a metaphorical bridge that connected the sublunary to the celestial realm and allowed the mind to see the flux and confusion of the everyday world as the mere illusory appearance of an underlying invisible ideal world.[14] It was geometry that revealed the ideal sphere within the rough rock or the straight line behind the farmer's wavering furrow. Through mathematics Plato came to understand the world as an imperfect image of the eternal realm, but he never conceived of mathematics itself as anything more than a metaphorical or analogical tool to aid understanding and facilitate enlightenment.

Beginning with Copernicus's sun-centered model of the Solar System, however, mathematical laws were treated

---

12. See T. S. Kuhn, *The Copernican Revolution* (Cambridge: Harvard University Press, 1957), 80. See also H. Butterfield, *The Origins of Modern Science 1300–1800* (New York: Free Press, 1965), 32.
13. See Barfield, *Saving the Appearances*, 49–50.
14. See Kuhn, *The Copernican Revolution*, 128.

as *actual descriptions* of nature. This view of mathematics as a literal, rather than metaphorical, description of nature represents the deepest and most radical influence of the revolution in thinking brought about by Copernicus, Galileo, and Newton. Mathematics was now a formal description of the actual mechanism behind all natural phenomena. This grand mathematical machine, like the simple mechanical devices it imitated, was mindless and inanimate, acting without meaning or purpose.

"MICROMECHANISTICS"

Despite radically new concepts, the power of the Newtonian system has continued to exert its influence in physics right into the current era. In the course of the nineteenth century, the mechanistic method was applied with modest success to the microlevel within matter to explain some of its bulk properties. The pressure and temperature of a gas, for example, were interpreted in terms of the average impact and speed of the countless molecules that make up the gas.

The promise of this scheme—to reduce the large-scale characteristics and behavior of bulk matter to the mechanical actions of its elementary constituents—was largely fulfilled in the twentieth century with the advent of quantum mechanics. The "mechanics" of quanta is quite different from Newton's. We no longer deal with material particles—atoms and molecules—as little hard balls, but instead we use a rather abstract mathematical description of the ultimate constituents of matter, which gives us no picture of electrons and atoms or any feel for what they are.[15] Whatever the ultimate elements of matter may be, they are still inanimate entities that obey mathematical laws, albeit in a random, probabilistic way.

Even the Solar-System model made it into early twentieth-century atomic physics. In 1911, Ernest Rutherford

---

15. See Chapters 8 and 9.

used Newton's model of the planets revolving around an attracting sun to describe how the electrons in an atom orbit an attracting nucleus. Although the atom is one hundred million trillion times smaller than the Solar System, this "planetary model" of the atom managed to rationalize the observed deflection of alpha particles in a beam passing through a thin gold foil. Niels Bohr would go on to modify and improve upon Rutherford's planetary model by introducing an atomic quantum theory, as we shall see in Chapters 8 and 9. Even in today's abstract and unpicturable quantum atom, electrons are still held captive by the attractive electrostatic force of the nucleus, and thus mimic Newton's gravitational model of the Solar System.

The clanking gears and wheels of an earlier age have been replaced by the more abstract, mathematical mechanisms of quantum theory, and the new machine of the microworld has indeed had spectacular success. The modern theories of atoms and molecules, however, are still of the nature of "dashboard" knowledge rather than "engine" knowledge. We have mathematical theories that can describe and predict the behavior of matter and its elementary constituents with incredible accuracy, theories that have given us semiconductors, microchips, and superconductivity. Yet we know no more of the whys and wherefores of the microworld than a novice driver knows of what goes on under the hood of a car. Our knowledge of things remains just as mechanistic, incomprehensible, and counterintuitive as Newton's original explanation of motion in terms of the mysterious force of gravity.

THE MACHINE RUNS DOWN

If the whole world is made of machines, then sooner or later everything must come to a grinding halt. Although idealized machines hypothetically would operate forever, real machines eventually wear out. Even in the best and most efficiently designed machines, friction is always present, heat is generated, and parts wear out. The job of machines is to convert energy into useful work. In the process, some energy is always wasted and given off as heat: no machine operates at 100 percent efficiency. The behavior of real machines cannot be described by the laws of mechanics alone. The wearing out of machines and the degradation of energy are both examples of a universal law—the second law of thermodynamics—that proclaims the irreversibility and inefficiency of all processes that use and transform energy. Because machines are big consumers of energy, it is inevitable that the subject of thermodynamics should be very relevant to a mechanistic model of the world.

An automobile engine, for example, converts about one-third of the chemical energy in the gasoline into useful mechanical energy for running the car. The other two-thirds is transformed primarily into heat in the car's cooling and exhaust systems. The second law of thermodynamics tells us that this is not the exception but the rule. Some energy is always degraded (i.e., converted into a form unavailable for doing work) in all machines and in all processes.

Our Solar System may look like it will run forever, but, in fact, it cannot. When the mechanistic view is applied to the universe at large, the second law predicts that the whole cosmic machine must eventually run out of steam—must wear itself out. All of the original energy of the universe will still be there, but it will have become unavailable for useful work, like the exhaust heat from a car. The universe will have run out of the energy it needs for all forms of motion and activity. In fact, the second law predicts the so-called *heat death* of the universe. In this ultimate, inevitable state, all processes and activity will come to a halt and matter and energy will reach a state of quiescence and inertness. Nothing, quite literally, will happen any more. The universe will have reached a state of minimum order and activity. The final consequence of a mechanistic universe is the degeneracy of all matter and energy into a kind of permanent state that is chaotic, random, featureless, and meaningless—not a bad description of hell, really.

The mechanical worldview rejects any idea of purpose in cosmic law. The universe operates according to meaningless laws of causality, which themselves are little more than an abstract mechanics of space and time. Ultimately the machine universe degenerates into a final state that is just as meaningless as its existence and operation. Is there no escape from this mechanistic fate?

CHAPTER 6

Cosmology

WE HUMAN BEINGS HAVE ALWAYS struggled to make sense of our existence. Since earliest times we have tried to use religion, myth, and philosophy to fathom the origin and evolution of the cosmos. More recently science has taken over the task of rationalizing the universe. (Was it really a *rational* explanation we wanted?) The twentieth century has seen a revolution in cosmology, as well as the revolutions in relativity and quantum theory. We no longer believe that the universe is static but that it is expanding from a primordial explosion—the big bang—that occurred fifteen or twenty billion years ago. In its breakneck expansion, the cosmos gives rise to stars, galaxies, pulsars, supernovas, and such exotic things as black holes, incredibly remote and radiant objects called quasars, and traveling waves—like ocean waves—in the gravitational field. In the past few decades, cosmologists have also brokered an unlikely wedding between the large-scale science of the universe and the microscopic science of atoms, protons, and their most elementary constituents—quarks. It seems we need both the quasar and the quark to decipher the big bang.

107

SURPRISING EXPANSION

During the first few decades of the twentieth century, astronomers discovered that the universe was much larger than anyone had ever realized. Galaxies, or great island conglomerates of stars, were found to lie at remote distances, far beyond the confines of our own Milky Way galaxy. Even more surprising than the enormous size of the universe was the realization that the universe was expanding. The vast cosmic environs of the earth were not static and fixed in size, as had long been assumed by Western civilizations, but were gradually increasing in size.

At first astronomers had observed that a number of the distant galaxies were receding, or moving away from us, at great speeds. This was determined by examining the spectrum of the light from these galaxies. If a light source is moving away from you, the frequencies of the various color components of the light will be lowered, or shifted toward the red end of the spectrum.[1] From the amount of the red shift, the speed of the source away from you can be determined. On the other hand, if the source is moving toward you, its light will be shifted toward the blue.

For the great majority of the galaxies analyzed between 1912 and 1925, it was discovered that their light was shifted toward the red. Most galaxies were moving away from us. In 1929, an American astronomer, Edwin Hubble, had succeeded in comparing his estimates for the distances of remote galaxies to their speeds. His surprising conclusion was that the more distant a galaxy was from us, the faster it was moving away. It was as if the whole universe was expanding and therefore every part of it appeared to be receding.[2]

---

1. In the spectrum of visible light, red light has the lowest frequency and blue the highest. If, as the result of relative motion, the frequency of electromagnetic waves (whether visible or not) is lowered, we call this a red shift. Correspondingly, a blue shift refers to a raising of the frequency.

2. See M. Rowan-Robinson, *The Cosmological Distance Ladder* (New York: W. H. Freeman, 1985), Chapter 1. This book gives an excellent account of how astronomical and cosmological distances are determined.

Imagine that you are on a rubber sheet surrounded by other people. Three feet from you is Dorothea, and three feet beyond her is Joshua, who is six feet from you. The rubber sheet begins to stretch. After a few seconds, Dorothea, who was originally three feet from you, will be six feet away. But Joshua, who was six feet from you, will now be twelve feet away. Joshua has moved twice as far as Dorothea in the same time. So Joshua, who was farther away from you, is moving away faster than Dorothea, who was closer. To Dorothea, you and Joshua are each receding at the same speed. And to Joshua, you are moving away faster than Dorothea.

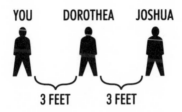

Now suppose that the rubber sheet you are on is actually the surface of a great balloon, which is expanding. The speed-distance argument still holds: Those who are distant from you will recede faster than those close to you. Everyone on the expanding surface will observe the same thing. Everyone on the surface will see everyone else moving away. No one will be at rest. No one will be at the center of the expansion.

This pretty much describes the expansion that Hubble attributed to the universe at large—although we must keep in mind that the two-dimensional rubber surface is only an analogy to the three-dimensional space of our physical universe. We must imagine ourselves on the three-dimensional "surface" of an expanding four-dimensional "balloon."[3]

Remarkably enough, Einstein's general theory of rela-

_____

3. For help with these dimensional gymnastics, you are referred once again to E. A. Abbott, *Flatland* (New York: Dover, 1952).

tivity had actually predicted a general expansion or contraction of the universe. Certain hypothetical models of the universe were analyzed by Einstein and Willem de Sitter in 1916 and 1917, and an expanding version was found. To "correct" this nonstatic universe Einstein committed the "greatest blunder" of his life, as has been mentioned, by adding the problematic cosmological constant to his equations.

HOW BIG IS THE UNIVERSE?

Today the expansion of the universe is accepted as an empirical fact by all scientists (not to mention the rest of us, who have very little choice in the matter). The most generally accepted explanation is the big bang theory, which conjectures the occurrence of an incredible explosion in the remote past that has expanded over the eons into the present universe with its vast proportions. "Just how vast?" you ask. The answer: pretty damn vast!

From its current rate of expansion, it is possible to estimate how long the universe has been growing. There is still a heated debate over this estimate, but it is in the neighborhood of ten to twenty billion years.[4] Because speed increases with distance, the farthest reaches of the universe move the fastest. At the remotest distances, the visible "edge" of the universe recedes from us at nearly the speed of light.[5] In fifteen billion years (we'll use fifteen billion as an average estimate), the edge of the universe has receded a distance of fifteen billion light-years. A light-year is a unit of distance, not of time. It is the distance light travels in one year. A light-year is about six trillion miles— roughly eight hundred times the diameter of Pluto's orbit in the solar system. The nearest star, besides our sun, is Prox-

---

4. See Rowan-Robinson, *The Cosmological Distance Ladder*.
5. In effect, this defines the boundary of the physical universe. If any region of space beyond this is "stretching" faster than the speed of light, then no electromagnetic signals, which are limited to the speed of light, could ever reach us from such regions.

ima Centauri, which is about four light-years, or twenty-four trillion miles, away. Our Milky Way galaxy, with its two hundred billion stars, has a diameter of about one hundred thousand light-years. The Andromeda galaxy, the nearest major galaxy to ours, is two million light-years away.

But these numbers are peanuts compared to the dimensions of the whole cosmos, which has an estimated diameter of thirty billion light-years. That's 30 billion times 6 trillion, or 180 sextillion miles. (Such numbers may help you; they don't do an awful lot for me. But there they are, for whatever they're worth.) So the universe is thirty billion light-years across and fifteen billion years old . . . and growing.

Cosmology and General Relativity

In Einstein's general theory, the matter content of space determines its curvature. Near the sun, space is more sharply curved, and so the inner planets revolve in more tightly curved orbits. Instead of talking of gravitational fields around large bodies, we speak of the curvature of space in their vicinity. The curvature of space in our Solar System or in the neighborhood of a star or planet is referred to as a local curvature: it characterizes the geometry of a confined or localized region of space.

By contrast, we can speak of the global or large-scale curvature of the entire cosmos. In this case, we are not interested in the localized "dimples" and distortions of space but in the overall shape and curvature of the geometry of the universe as a whole. Sooner or later someone was bound to ask what the global curvature is of the universe at large. Einstein had already considered this question in his earliest papers on general relativity. As it turns out, the global curvature of cosmic space-time is related to the rate of expansion of the universe.

The connection between the curvature of space and the rate of expansion of the universe is made in the equations of general relativity. It's a little easier to understand this connection by using the old language of gravitational fields. The big bang represents a phenomenal explosion whose blast sent the matter and energy of the universe spewing out in all directions.[6] Since there exists a gravitational attraction between all material bodies, then every particle of matter in the explosion is attracting or pulling every other particle, and this slows down the expansion.[7]

If the mutual attraction of all the matter is strong enough, it can overcome the expansion caused by the original explosive force of the big bang. In this case, the universe would stop expanding at some maximum size, and then it would begin to contract under the mutual gravitational attraction of all the cosmic matter. This depends on two things—how much matter there is in the universe and how the matter is distributed or spread out in space— because the strength of the gravitational attraction between any two bodies depends on their masses and also on the distance between them. So the rate of expansion of the universe is determined not only by the total amount of cosmic matter but also by the distribution or concentration of that matter. The more concentrated the matter is, the more it is slowed down by the mutual attraction. If the concentration of matter is great enough, then all the mutual attraction will ultimately bring the expansion to a halt and then cause the universe to contract. If, on the other hand, the concentration is too low, then there will not be enough gravitational attraction to stop the expansion, and it will go on forever.

---

6. Incidentally, this is not an explosion *in* space-time, but *of* space-time! This is very difficult to understand or visualize, but all of space and time were created in the big bang, along with all of the matter and energy of the universe. (We have already seen that general relativity does not really distinguish between matter and space-time.) So one can speak neither of the cosmos expanding in space nor of a time before the big bang.

7. We use the term *matter* here to refer to both matter and energy, which are equivalent in relativity.

Thus there are two possible scenarios in the evolution of the universe: The cosmos will eventually reach a maximum size and then contract toward what cosmologists whimsically refer to as the big crunch. Or the cosmos will expand forever. According to Einstein's equations, these two scenarios, in turn, correspond to two different kinds of cosmic geometry or global curvature.

We've already seen that gravity is reinterpreted in Einstein's theory as the local curvature of space. By the same token, the mutual gravitational attraction of all of cosmic matter (or alternatively, its concentration) is reinterpreted in general relativity as the global curvature or geometry of the whole universe. In the big crunch scenario—the case of too much gravity, so to speak—cosmic space-time turns out to be curved in four dimensions like the surface of a sphere in two dimensions. The cosmos is then said to be "closed," which is analogous to the geometry of a sphere whose surface closes on itself. In the other case of too little gravity, space-time turns out to be curved like an hourglass or a saddle-shaped surface. In this case, cosmic geometry is said to be "open," which is analogous to saddle surfaces that flare out and never close on themselves.[8]

It turns out that there is an in-between case of the concentration, or density, of cosmic matter in which cosmic geometry is neither open (spherical) nor closed (saddlelike) but is flat, which is analogous to a two-dimensional plane. Gravity is neither too strong, nor too weak, but just right. The universe would still expand forever, but it would come to a halt exactly after an eternity of time (as opposed to still going at a good clip when it reaches eternity!). This case is extraordinarily improbable. Think, for example, of the infinitely many possible spheres that can represent closed surfaces and of the infinitely many saddles that can represent open surfaces. Then imagine the one and only flat plane that constitutes the case between the

---

8. See the discussion of the geometry of spheres and saddles in Chapter 3, footnote 6.

infinite number of closed and open surfaces. One case out of infinitely many is all but impossible. This unlikeliest of cases wouldn't be worth mentioning but for the fact that it seems to be the most accurate description of our very own cosmos!

WHAT'S THE MATTER?

Is the universe open, closed, or flat? There's indirect evidence and some theoretical arguments to suggest that the universe is flat.[9] But to find out for sure, we need hard data. We know that the universe is expanding, but we cannot tell the rate at which the expansion is slowing down. So, if we can't measure the rate directly, we must measure the curvature, i.e., the concentration of matter. Concentration is determined by how much matter there is and how large a space it fills. The more matter there is in a given space, the greater the concentration. We know the size of the universe. So we must measure how much matter there is in the universe. This turns out to be quite impossible.

The trouble is you can't tell whether you are seeing all the matter there is. And seeing is the right word, because all the matter we know about in the universe is visible or luminous matter—stars, galaxies, glowing gas, reflecting planets. How much is that? If we count up all the visible stars and galaxies, we find that the density or concentration of luminous matter is about 1 percent of the density required for a flat universe. In other words, the mutual gravitational attraction of all the luminous matter in the universe is not enough to ever stop the expansion. The universe is open.

We know though that there is more than luminous matter. There are cold stars that do not glow. There is gas and dust that is not illuminated. And there is matter whose existence we can infer from the motion of galaxies.

9. See L. M. Krauss, "Dark Matter in the Universe," *Scientific American* (December 1986): 58.

The Milky Way is like a great disk of stars and gas rotating as a unit in space. If the Milky Way contained nothing but the luminous matter we see, it would not rotate as a unit—the outer portions of the galaxy would trail and drag behind the more rapidly rotating inner part. Therefore, we conclude that the Milky Way must contain matter we cannot see or directly detect in any way. We know it's there because of the mechanics of rotation. Similar arguments can be made about most galaxies in the cosmos and even about the large-scale motion of galaxies in gigantic clusters and superclusters.

By this indirect method of reckoning, we can infer a cosmic density that is roughly 20 percent of the critical density required for a flat universe. That's twenty times the density of luminous matter, but it still leaves us in an open saddle-shaped universe that lacks 80 percent of the matter needed to flatten it. Where is that matter and what is it? Since it's invisible and undetectable, it's called dark matter. But whether it actually exists, where it is, and what form it takes remain great mysteries of modern cosmology.[10] If the big bang is a correct theory, and if the universe is as flat as we think, then for thousands of years we have been preoccupied with a mere 1 percent of nature.

TO BANG OR NOT TO BANG

Why then is the big bang the prevailing theory of the day? It has some serious flaws and presents nagging problems, which, if anything, have gotten worse in the past decade. But despite this, the lack of any serious contenders to the big bang theory makes cosmologists, like all of us, prefer, as Hamlet said, to "bear those ills we have than fly to others that we know not of." It was not always so, however. In the fifties and sixties there were two theories vying for the right to explain the cosmic expansion. The big bang had its arch rival—the steady state theory. This theory

10. See Krauss, "Dark Matter in the Universe."

proposed a universe that was steady or unchanging, except for its size and the amount of matter it contains. In other words, in the steady state theory, there is a continual creation of matter out of nothing, causing the universe to expand as the amount of matter increases.[11] The rate at which matter is created is exceedingly small—an atom per century per 1,000 cubic miles of space (nothing you'd really notice around the house). It does add up, though, in fifteen billion years.

The newly created matter, according to the theory, always takes the same old familiar form—stars, galaxies, planets, and so forth. The universe doesn't evolve, as such; it simply creates more and more of the same thing. And although this scenario has a certain appealing simplicity, it was the absence of change and evolution in the steady state theory that ultimately did it in. It's all old hat by now, but it's still worth considering how the big bang won the day because it will reveal the difficulties that any cosmological theory must face. We'll briefly consider the five major arguments favoring the big bang.

1. *Evolution:*[12] As astronomers look around the universe, they see great variety among the stars and galaxies. Many galaxies have a well-defined symmetric spiral form. Many others are rather amorphous and irregular. And there are many with a character somewhere between the two extremes. The story is similar for stars, which display a great range of size, color, brightness, and stability. One can simply accept such a variety as the natural state of affairs, or one can try to make some sense of it. Astronomers have had good success (more so with stars than with galaxies) in interpreting the observed variety as the stages of evolution of stars and galaxies.

---

11. If the creation of matter out of nothing seems an unlikely hypothesis, it's no less credible than the sudden and instantaneous appearance, at the moment of the big bang, of all the matter in the universe.
12. For an excellent treatment of the evolution of matter in all its forms, physical and biological, see M. Taube, *Evolution of Matter and Energy* (New York: Springer-Verlag, 1985).

Stars are born as cosmic gas and dust coalesce under their own gravitational attraction (a kind of "small crunch"). In this condensation, the star heats up. Eventually, it reaches a temperature high enough to ignite the nuclear fusion reaction in which hydrogen is fused into helium and enormous amounts of energy are released.[13] The star lives most of its life in this state, producing great quantities of light and heat for millions and even billions of years. Our sun, for example, is about halfway through its projected ten-billion-year lifetime, and we are the fortunate recipients of just a tiny fraction of its radiant energy.

But all good things must come to an end. The star finally runs out of hydrogen fuel, begins to collapse once again, heats up even more, and then ignites a helium-to-carbon fusion reaction. This induces a gargantuan explosion that expands the outer layers of the star. In the course of the expansion, the outer layers cool from white-hot to merely red-hot. This is the so-called red-giant stage of stellar evolution. When our sun reaches this stage in five billion years or so, it will expand beyond the earth's orbit, destroying all of the inner planets in the process.

The scenario after the red-giant stage depends on the size of the star. A star somewhat larger than the sun will undergo a series of cataclysmic events, culminating in a spectacular supernova explosion, such as was witnessed in February 1987.[14] After this, the large star has no further possibility of fusion (it uses up just about every nuclear reaction in the book during the supernova explosion). It will condense into a spinning neutron star—a pulsar—and finally perhaps into that strange anomaly of general relativity and cosmology, a black hole.

A smaller star, like the sun (and stars can be a good deal smaller than the sun, which is about average in size) experiences a much less dramatic ending. It becomes

13. For a discussion of nuclear fusion, see Chapter 16.
14. See S. P. Maran, "A Blue Supergiant Dissects Itself in a Cosmic Explosion," *Smithsonian* (April 1988).

highly compact and condensed and passes through a series of stages—white dwarf, red dwarf, brown dwarf—and darkens in color as it slowly cools with the passing eons.

All of these stages have been studied in great detail, and they fit very well into a model that rationalizes the stages of stellar evolution in terms of known nuclear reactions and gravitational attraction. Such an evolving scenario is very difficult to explain in a steady state model, but it agrees quite well with the big bang picture of a dynamically evolving cosmos. In the steady state theory, which assumes an unchanging universe, you must assume that stars are randomly created in these various stages, with no rhyme or reason. Without the time sequence of stellar evolution, and the supporting corroborations of nuclear physics, it becomes very difficult to make sense of the variety of stars.

Galaxies are much less well understood, but most astronomers are convinced that the variations among galaxies also reflect the stages of galactic evolution. Thus the evolving changing cosmic picture of the big bang theory seems more in keeping with the dynamic volatile universe we observe.

2. *Quasars:* During the 1960s, objects were identified in the skies that have very puzzling properties. They are comparable in size to stars but are located only at the remotest distances from us. And yet they radiate enormous amounts of energy—often a good deal more than the average galaxy. Because of their relatively small size, they were called quasi-stellar objects, which was eventually shortened to quasars. How could such small objects emit such huge quantities of energy? In ordinary galaxies, the radiant energy comes primarily from the individual stars. Yet a quasar, thousands of times smaller than a galaxy, often emits ten times as much energy. Clearly some new source of energy must be responsible for this phenomenon.

Much is still unknown about quasars and their abnormal output. Astronomers have come to believe that quasars are actually the cores of extremely active galaxies that release great quantities of gravitational energy when gas and dust fall into the maw at their very compact and massive centers. Most cosmologists agree that they are found only in regions that are exceedingly remote from us in space, and therefore also very remote from us in time.[15] Once again, it is much easier to understand quasars from the big bang viewpoint. They can be associated with some earlier more dynamic stage of cosmic evolution, which precedes the present by many billions of years. We see no quasars in our cosmic space-time neighborhood presumably because we are long past the quasar stage of cosmic evolution.

A proponent of the steady state model, on the other hand, is forced to postulate arbitrarily that quasars are created only in remote regions of the galaxy. But why would that be true if the cosmos and its ongoing creation are supposed to be uniform?

3. The Cosmic Background: In 1965, Arno A. Penzias and Robert W. Wilson discovered a previously unrecognized form of cosmic radiation that bathed the earth uniformly from all directions of space. They realized that this radiation did not originate within the earth's atmosphere— or even within our galaxy—because radiation from those sources could not possibly have been so uniform. Instead, they recognized it as a kind of "afterglow" from the primeval expanding fireball. This radiation had been predicted by George Gamow in 1953 as a consequence of his big bang hypothesis. From the spectrum of this cosmic background radiation, it was possible to determine its temperature, which was found to be slightly less than 3° above absolute zero (about −455°F). This very low temperature represented the cooling of the original fireball after some fifteen billion years.

In 1978, Penzias and Wilson were awarded the Nobel prize in physics for their discovery, which is one of the

---

15. Recall that light from the most distant bodies in the cosmos takes billions of years to reach us.

most celebrated in modern cosmology. It also became the largest nail in the coffin of the steady state theory. There is no steady state rationale or explanation for the universal cosmic background radiation.

4. The Makeup of Cosmic Matter: Another fatal blow to the steady state theory is the explanation of the relative abundances of the chemical elements in the universe. Of all cosmic matter 92 percent is hydrogen, 7.8 percent is helium, and the remaining 0.2 percent is everything else. How and why did it get that way? The answer is a long story (which we'll try to shorten), involving the evolution of matter according to the big bang scenario. One starts with the estimated temperature, pressure, and density of the primordial matter in the big bang and then calculates the sequence in which various kinds of matter are formed as the universe expands and cools. The most exotic particles, such as quarks, form early on and then combine into protons and neutrons. All of this occurs within the first second of the lifetime of the cosmos! As the cosmos cools further over the next hundred thousand years or so, neutrons and protons first aggregate into nuclei, which later combine with electrons to form atoms, starting from the simplest atom—hydrogen. A few basic elements form at this stage—deuterium (an isotope of hydrogen), helium, and lithium. The higher elements are produced at a much later stage in cosmic evolution—in the supernova explosion of large stars.

The calculated percentages of the chemical elements, particularly of those formed earliest—hydrogen, deuterium, helium, and lithium—agree very well with the observed abundances.[16] The steady state theory can do nothing but postulate arbitrarily that the chemical elements are created from the outset in the observed abundances.

5. Unification: The discovery of the cosmic background radiation together with the successful calculation

---

16. See D. N. Schramm and G. Steigman, "Particle Accelerators Test Cosmological Theory," *Scientific American* (June 1988).

of the observed abundances of the elements were the death knell of the steady state theory. The big bang theory is far from an established fact. (It's hard to imagine how any theory about events occurring some fifteen billion years ago ever could be.) But its rational explanation for many experiments and its agreement with many observations make it the most satisfactory and unifying theory so far.

It also has another unique advantage in the history of physics. The big bang scenario makes use of the latest theories of elementary particles in explaining the evolution of matter. These new theories assume that the complexity of the contemporary universe is the end product of the much simpler and unified state of the early cosmos. Instead of the four forces of nature that we find today—gravity, electromagnetism, and the strong and weak nuclear forces—presumably there was only a single unified force in the very earliest instants of the big bang. The new theories also assume that the elementary particles of today were combined in some simpler unitary form fifteen billion years ago. Such conjectural ideas are formalized in today's unification theories, about which we'll have more to say in Chapter 18. But although one conjectural theory can hardly prove another, this unlikely marriage of the macro- and microworlds tempts the witnesses and wedding guests to have greater faith in the union than in the separate partners. How stable and enduring the marriage will be remains to be seen.

New Answers, Old Questions

In the spring of 1992, a major new discovery was announced that adds renewed credibility to the big bang theory. Data from NASA's Cosmic Background Explorer satellite (COBE) indicated for the first time that there are minute fluctuations in the temperature of the cosmic radiation, which is the afterglow from the big bang fireball.[17]

17. See M. Stroh, "COBE Causes Big Bang in Cosmology," *Science News* (2 May 1992): 292.

Previously, this background radiation had been found to be uniform to a very high degree. In other words, the temperature of different portions of the radiation had been measured essentially to be the same. This radiation has been traveling through space since the universe was a mere three hundred thousand years old, which means the radiation is almost as old as the whole cosmos—fifteen to twenty billion years. The trouble with this primeval uniformity is that it was inconsistent with the highly nonuniform character of today's universe.

Everywhere there are great inhomogeneous island galaxies containing countless dense stars separated by empty space or gases and dust. And the spatial distribution of these galaxies is anything but uniform. Great empty voids of cosmic space, hundreds of millions of light-years across, are "bounded" by bubble- and foamlike stretches of galaxies. Clusters and superclusters of galaxies are the rule rather than the exception. In one case, a monstrous "block" of galaxies, called the great wall, is five hundred million light-years long. How could these titanic nonuniform structures have formed from the innate smoothness of the early universe?

The recent COBE data suggest that the newly discovered minute temperature variations in the cosmic radiation are actually the ancestral "seeds" of inhomogeneity in the primal matter and energy, which ultimately grew into today's stars and galaxies. As the radiation expanded and cooled, the slightly hotter and cooler spots gave rise to variations in the density of matter, which gradually condensed into "lumps" because of the force of gravity. Wherever the material in space was slightly more concentrated, its self-gravitational attraction caused it to condense further and finally "jell" into dense galaxies and stars, leaving empty space around it. Cosmologists see the new data as the "missing link" that fills the gap between the early, almost uniform big bang and today's lumpy and irregular pattern of voids and galactic superclusters.

And yet these minute temperature fluctuations by

themselves may not be adequate to explain how today's cosmic structures formed—even in a time span of fifteen billion years. Other unknown mechanisms may have been involved; so the theory is not yet capable of tracing cosmic evolution in detail.

Other serious questions remain: The geometry of the universe appears to be flat, as we discussed earlier. There is no astronomical or other evidence of any large-scale global curvature of cosmic space.[18] There are good theoretical reasons to believe that the universe is flat. But the density of the observable luminous matter is only about 1 percent of the critical density required for a flat universe. We can infer about 20 percent by other means, and some cosmologists believe there must exist exotic invisible forms of "dark matter" throughout the universe to make up the other 80 percent of the critical density. But the fact remains that even at 1 percent of the critical density, the universe is incredibly close to being flat. The big bang scenario predicts that slight departures from flatness in the early cosmos would have increased enormously with time. So the fact that today's universe is nearly (if not precisely) flat implies that in its early infancy the big bang must have put even the flattest of pancakes to shame.

Of all the possibilities of global curvature that general relativity allows for an expanding universe, it is almost inconceivable that ours should be flat. It seems that Euclid was right after all! The chief contending theory today to explain our extraordinary flatness is called the "inflationary" model of the big bang, which postulates a sudden rapid expansion of the cosmos during the first instants of its existence. This inflationary expansion can account for the flatness and large-scale smoothness of the universe, and, into the bargain, it also provides a conjectured mechanism for the creation of all matter and energy from the primordial vacuum. (This is no modest theory.)

---

18. See L. Abbott, "The Mystery of the Cosmological Constant," *Scientific American* (May 1988).

Yet even with inflation, dark matter, the new COBE temperature fluctuations, and other assorted exotic variations on the big bang, we've yet to explain the origin and evolution of galaxies and distant quasars or to fathom the events of the first fleeting instants of the big bang. And, of course, we still haven't the foggiest about what started the whole big bang process in the first place, and why, among all the myriad possibilities, it took this particular form.[19]

19. For a brief survey of the controversy, see I. Peterson, "State of the Universe," *Science News* (13 April 1991): 232.

CHAPTER 7

Science vs. Religion

IN OUR SURVEY OF COSMOLOGY in Chapter 6, we touched
on issues that inevitably conflict with traditional ideas in
religion. How did the universe originate? What laws gov-
ern its behavior and evolution? How did life come about?
Are there other forms of intelligent life in the cosmos that
have alternative versions of science and religion? Conflict
and antagonism between matters scientific and spiritual
are neither uncommon nor unusual. The origin and evolution of life are dealt with in both biology and the Bible, the
nature of mind in both psychology and Eastern scripture,
and the laws of nature in both physics and mythology.
These territorial disputes are no accident. Religion and
science have a common origin in the ancient human quest
to find meaning and purpose in life.[1] Only fairly recently
in history was the separation clearly made between the
empirical knowledge of nature and the spiritual under-

---

1. See J. Needleman, *A Sense of the Cosmos* (New York: Dutton, 1976).
For an excellent discussion of issues in science and religion, see I.
Barbour, *Religion in an Age of Science* (San Francisco: Harper & Row,
1990).

standing of existence. Science credits much of its great success to this separation. The final judgment has yet to be made on the legitimacy of the separation and the price of the success.

THE ETERNAL QUEST

The ancient shepherds who often lay under the brilliant night sky couldn't help but wonder about the beauty and order of the heavenly splendor. The musings and wonderment of early stargazers, stimulated by the natural human desire to make sense of existence and the world we inhabit, ultimately developed into astrology and astronomy. These early efforts to understand things were broad and all-encompassing. They combined forms of knowledge that we separate today into myth, philosophy, religion, and science. The mixing in astrology of what we now classify as astronomy and psychology seems peculiar and unacceptable to the modern scientific mind, but it's not hard to understand from the ancient point of view.

Early watchers of the skies were not motivated simply to describe the motions of the stars and planets, although this was certainly a part of their aim. They sought meaning and purpose in all of existence, and, therefore, in nature. The order and harmony of the heavens was especially suggestive of mind and intelligence in the cosmos, which is reflected clearly in Platonic philosophy and Eastern religion.[2] The very idea of treating stars and planets as purely physical inanimate bodies, which seems so obvious to us, would have been inconceivable to the ancient mind. For Plato, the cosmos was an organic whole—a living being animated by mind, intelligence, and spirit. The heavens were a reflection of the perfect order and harmony of the highest levels of mind and beauty—the realm of

2. See, for example, G. F. Parker, *A Short Account of Greek Philosophy* (New York: Harper & Row, 1967); Barfield, *Saving the Appearances*; and A. Daniélou, *The Gods of India* (New York: Inner Traditions, Ltd., 1985).

Ideas. The contemplation and understanding of this realm was both the intellectual and spiritual goal of human existence. The highest good and the greatest beauty were inextricably intertwined for Plato. He sought answers to the ethical problems of life in the aesthetics of the cosmos. For him, understanding the pattern and behavior of the stars was not merely a "science" in the modern sense of the term but a spiritual quest—a search for understanding and meaning, the pursuit of wisdom and bliss. The connection between the heavens and the human psyche was perfectly natural and logical to the ancient mind.

Because the Greek cosmos was a realm of wisdom and intelligence, it was to be studied and interpreted for meaning, purpose, and value. Not only the heavens but all aspects of the natural world served as a spiritual guide. Today we criticize as primitive and anthropomorphic the nature religions of early times. To reject them as superstitious, irrational, and naive, however, is to beg the whole question of human knowledge and wisdom. It is to assume that the modern tendency to compartmentalize knowledge, to divide spirit from matter, and to divorce religion from science is somehow permanently ordained in thought and forever beyond question and doubt.

In the medieval world, we find the search for meaning in nature reflected in alchemy. Once again, we are too glib about an earlier science (or pseudoscience, as we prefer to call it). We tend to reject alchemy as a strange and meaningless mixture of magic and religion, or (at best) an embryonic form of chemistry. But alchemy, as Jung and others point out, was not simply a foolish quest to transform base metal into noble gold. It was an effort to purify the ignoble and imperfect human soul and raise it to its highest and noblest state. It was a religious quest—not a purely scientific one. Matter and spirit were inseparable to medieval alchemists, and so they strove to transform matter and spirit through the same procedures, which amounted to sacramental rites and religious ritual as much as scientific research.

By contrast, in the modern era scientific knowledge is required to be objective, logical, empirical, and quantitative. Modern science divests itself of any subjective taint, value judgment, or sense of meaning and purpose. The teleological stance of Aristotelian science, for instance, is thoroughly rejected today as naive and anthropomorphic. It was the striving after final causes and ultimate purpose and meaning that diluted and confused the Greek system and other early systems of scientific thought. Only with the rejection of teleology and anthropomorphism was science able to make any genuine progress, to which the past three centuries of Western scientific achievement bear brilliant testimony.

Religion, on the other hand, dwells fundamentally and primarily in the domain of meaning. Its methods of gaining knowledge are matters of faith, contemplation, and revelation. It is deeply subjective, requiring the patient refinement of the human powers of intuition, imagination, and insight. It seeks the inner meaning, purpose, and value of existence. Religion is, on the face of it, the antithesis of science. It is always on issues of meaning and value that religion contrasts most profoundly with science.

THE STORY OF CREATION

Nowhere is the disparity between religion and science greater than in the description of the creation, to which religion attaches the deepest and most far-reaching significance and which science treats as a remarkable drama but without fundamental meaning or purpose. In this context, let us briefly review once more the main features of modern cosmology.

The cosmos known to science today is of a magnitude beyond the wildest imaginings of previous civilizations, not only of Greek scholars but even of nineteenth-century astronomers. As we discussed in Chapter 6, the distance from edge to edge of the visible universe is approximately 30 billion light-years, or 180 sextillion miles. This is all but inconceivable to the human mind. But however we may struggle to comprehend and appreciate 30 billion light-years, one thing is clear. The size of the universe totally dwarfs anything on the human scale, which by comparison is reduced to puniest significance. Contemplating nothing more than the vast physical size of the universe is enough to humble and intimidate us minuscule earthlings.

There are, of course, other important characteristics of our cosmos to consider. For instance, lightness, or rather darkness. We happen to live in the neighborhood of a reasonably intense light source (if you can call ninety-three million miles neighborly), but our situation is by no means typical. The universe consists of trillions of island galaxies, each containing, on the average, hundreds of billions of stars. We live in such a galaxy—the Milky Way. Despite these countless luminous stars and galaxies, most of the cosmos is empty space. The distances between the stars are enormous compared to the size of the stars. Our sun is nearly a million miles in diameter and is twenty-four trillion miles from the nearest star in its neighborhood. Far more vast are the spaces between the galaxies, which are typically separated from one another by distances that are tens and hundreds of times their own size. If you could beam yourself to some typical average locale in the cosmos, you would be millions of light-years from the nearest sources of light. You would, in fact, be surrounded by an oppressive inky blackness, pierced only here and there by the tiniest pinpoints of light, which would be the images not of stars but of galaxies that are so far away as to be all but invisible. Not a very inviting prospect for such sun- and light-worshipping creatures as ourselves.

What about heat? We observed in Chapter 6 that the average background temperature in outer space is three kelvins, i.e., three Celsius degrees above absolute zero (about −455°F)—the rock-bottom temperature for matter. In International Falls, Minnesota—the nation's icebox—temperatures often reach −40°F or −50°F in winter, extreme but livable temperatures for human beings, if they are well enough prepared for them. But winter in International Falls, or even at the South Pole, would be a tropical heat spell compared to outer space. At three kelvins, no form of life could possibly survive that even remotely resembles life on earth or any similar organic form of life that we might imagine.

The fact of the matter is that just on the basis of size, light, and heat, the current scientific picture of the cosmos could hardly present a more inhospitable, alien, and lethal environment for human beings. An inconceivable vastness that deflates our earth to less than insignificance, an impenetrable darkness that blinds and isolates us totally, and a frigid coldness that barely sustains activity, let alone life. Not a very inviting place to set up housekeeping.

But, obviously, the cosmos isn't all that bad or we wouldn't be here in the first place. It has an occasional, rare pocket of heat, light, and sustenance, and we are fortunate enough to inhabit one. We live on a tiny speck of dust, called planet Earth, which dizzily circles an average dwarf-type star called the Sun.[3] On this tiny planet, the forces and laws of nature have somehow conspired to bring about a stable and nourishing form of matter that has proceeded to transform itself into life and to consummate that life with consciousness. And here we stand, the shining end product of this remarkable transmutation—the most organized form of matter and the highest form of life on earth, very possibly in the whole universe. Indeed, we may be the only living things in the universe. Carl Sagan and his extrapolations notwithstanding, there is as yet not the slightest shred of evidence of life anywhere else.[4] Other life forms may exist or we may be entirely alone in the universe. It remains an open question. In any event, we see

---

3. The sun is viewed in astronomy as a typical star—approximately midway between the brightest and faintest stars. It is also classed as a dwarf star in terms of its most probable future evolution and fate.
4. See Carl Sagan, *Cosmos* (New York: Random House, 1980), 298-302.

ourselves as the crowning achievement of four billion years of evolution on earth.

We mustn't let any of this go to our heads however. The whole incredible big bang story with its climactic human ending is merely the result of physical and chemical processes that are completely random, accidental, and meaningless. We may think we're pretty good and important, but there is absolutely nothing in the blind, meaningless events to suggest the slightest purpose, value, or significance in our existence. If we should be so foolish as to annihilate ourselves in an atomic holocaust or through the strangling pollution of the earth, it will make not the least difference in the scheme of things. The planets, stars, and galaxies will continue on their cosmic schedules, completely oblivious to our passing. So much for human significance.

This picture we have been contemplating, however tragic, ironic, or even comical it may seem, is, in fact, a more or less straightforward account of current thinking in cosmology and physics. I may have exaggerated slightly here and there for dramatic effect, but what I've presented is essentially the straight party line on the cosmos. But it is something else, as well—something we never associate with science. It is our creation myth. After all, it is the story of how we got here, isn't it? If you object to calling it a myth because it is more factual than the word *myth* suggests, then you must realize that there are many essential episodes of this story that are decades—possibly centuries—away from the remotest possibility of verification. Many scientists will tell you that the big bang is alive and well, but that doesn't mean that the theory isn't filled with conjecture and extrapolation. It's simply the best theory we have for now.[5]

Worst of all is the effect—mostly unconscious—that this myth has on the human psyche. Suppose you try for a

---

5. See J. Silk, "The Big Bang Is Alive and Well," *Physics Today* (book review), (June 1991): 106.

moment to dream up your own alternative to our creation myth but with the following proviso: it must terrify and scare the living daylights out of everyone you tell it to. I doubt very much that anyone can do better than our current big bang myth. If taken emotionally and with full psychological awareness, this is a story that can fill the human heart with awesome terror. And yet, we are supposed to stiffen our upper lips and view this account of creation dispassionately, and even with a certain respectful appreciation, as a fabulous but utterly meaningless pageant.

SCIENTIFIC IDOLATRY

The title of this section—Scientific Idolatry—may seem a little surprising or puzzling. You may think of idolatry as a rather archaic term that, in any case, has no place in a discussion of science. But the term is not nearly as inappropriate or as anachronistic as it sounds.

In the Bible, idolatry refers to the false attribution of divine powers to humanly created artifacts or idols—for instance, treating the statue of a golden calf as if it were a god. The biblical sin of idolatry refers not simply to making "graven images" but to worshipping them. Idolaters are caught in an illusion—a kind of self-hypnosis. They disregard and repress the memory of their own role in creating an idol, and, consequently, they think of that idol as something external, mysterious, and divine. The idol, in effect, becomes a god and takes on powers far beyond those of the human beings who actually created it. Where does this power come from? Why, from the idol builders and worshippers themselves, who, by a kind of unconscious or hypnotic consensus, agree to bestow a divine authority on their statue and to ascribe to it an unlimited power over themselves.

You may be wondering what any of this has to do with science. Today, the average person—including a good many scientists—treat the ideas, concepts, and theories of science in exactly the same way as the ancients treated their golden calves. We take quarks, black holes, and the big bang story to be objective elements in an authoritative description of an external, independent reality. We forget or suppress the fact that all of these elements are ideas that came originally from the human mind, as do all the arguments we use to justify them. All scientific concepts and theories, together with the whole system and rationale of the so-called scientific method, clearly originated in the human mind. The complex, extensive, detailed, and astonishing picture of the natural world that we call science is the product of human imagination, thought, insight, and genius. It is neither external, nor independent, nor final, nor even provable. The whole structure and content of science, including its fabled empirical method, is like a vast and intricate game whose rules, playing board, and pieces were all created by human beings for their own use, benefit, amusement, power, and security.

It wasn't for nothing that the Old Testament prophets singled idolatry out for special condemnation. We can easily see its destructive effects on modern civilization. Every other culture in history has invented or "received" a creation myth whose express purpose was to rationalize human existence—to tell us who we are, how we got here, and what our value and purpose are. Indeed, it is the job of creation myths to tell us the meaning of life. Modern science does precisely the opposite. For the first time in history, a culture has conjured up a story about itself that altogether denies any meaning, value, or purpose in human existence. This is far more perverse than simply fouling our own nest; this is a total denial of any need for a home, a haven, or any sense of belonging. In place of a nurturing, participatory universe, we are supposed to soothe and pacify ourselves with a detachment and objectivity toward our environment which can only be considered pathological.

Is it any wonder that we befoul and trash the natural environment today for the sake of our technological imperative, that we lay waste to the world's great rain forests to advance agriculture and increase its profits, that we desecrate ancient burial grounds and hallowed native sites to further the study of humanity, that we perform hideous experiments on animals in the pursuit of biological and medical research, that we make scientific technology the indispensable handmaiden to a political system obsessed with power and control, and that we convert morality and aesthetics into abstractions in our insatiable quest for objectivity? The rejection of nature has become the hallmark of our civilization, which denies to itself any spirit or cosmic meaning.

Science as Our State Religion

Strangely enough, there are remarkable similarities between religion and science. Each is based on experience—religion on inner, science on outer experience. They both use specialized language for formal, precise communication—Latin, Hebrew, and Sanskrit are to religion as mathematics, computer programming, and electronic-circuit diagrams are to physics—in contrast to the vernacular, which is used for informal communication, especially with the lay public. Each has its revered body of literature in the form of scripture or research journals and monographs. Each requires a prolonged and austere period of technical training in monasteries and seminaries or in graduate schools and laboratories. The accomplished professional in each field commands a deep respect from the lay public and is treated as an unquestioned authority on matters of divine or natural law.

In a sense, science has taken over the role of state religion in modern culture, and it has become a very influential religion at that. Who can deny that the scientific establishment has become a modern priesthood? The pronouncements of scientists are respected and accepted by today's public just as the doctrines of the church fathers were respected and accepted by people a thousand years ago. The rigorous training in arcane mathematics and methodology is no less exacting, demanding, and monastic than was the medieval study of ancient languages and theology. Modern scientific training today is an insuperable barrier to the layperson who would question the authority of science, just as the ecclesiastical training of the Roman Catholic priesthood was a great obstacle to the medieval laity with its questions and doubts. If anything, modern science incurs far less challenge and criticism than the church ever did. The church fathers would have given their eyeteeth to command for medieval Catholicism the kind of obedience and blind faith that we freely lavish on science today.

The church long ago recognized the threat posed to it by science. Among the heretics tried by the Inquisition was Giordano Bruno, who was burned at the stake in 1600 for professing that the earth revolved about the sun. Even Galileo was put on trial for his views, forced to recant them, and then sentenced to remain a prisoner in his home for the final decade of his life. These and similar episodes are not proud ones in religious history. Although we condemn them today, we cannot fail to appreciate the church's efforts to defend itself in a war that it ultimately lost to science, just as it had feared. It is science and not religion that gives today's world its rationale, morality, sustenance, and story of creation, such as it is. The concessions of the church to science have gone so far that even in its most mysterious acts and pronouncements it dares not contradict the findings of science. In evaluating the authenticity of the Shroud of Turin or in canonizing a saint, the church leaves no stone unturned in its efforts to achieve scientific respectability.

Despite protestations that science has nothing to do with religious and spiritual questions, it is science that dictates to the church and not vice versa. It is science that determines the character and philosophy of our civilization. The danger in all of this is that it has happened unconsciously and without deliberation or consent. Science has been of enormous and indisputable material benefit to modern civilization; but if we fail to recognize and take account of the deeply religious role that science plays in our lives, we run the risk of destroying not only our material benefits but our souls as well.

CHAPTER 8

Photons and Electrons

IF THE TRANSMISSION OF LIGHT through space was the key to relativity, then it was the emission and absorption of light by matter that triggered the quantum-theory revolution. In 1900, Max Planck introduced his quantum hypothesis to explain the puzzling coloration of light emitted, for instance, by a hot metal object that can glow red or yellow or white depending on its temperature. Five years later, Einstein adapted Planck's ideas to explain the photoelectric effect in which absorbed light ejects electrons from the surface of a metal. Einstein assumed that tiny particles of light—photons—acted like speeding billiard balls that bounced the electrons out of the metal. The hypotheses of Planck and Einstein directly contradicted Maxwell's wave theory of light, but without them, no one could explain the radiation from hot metal or the photoelectric effect. Similar ideas were used later to decipher the radiation of light from atoms and the wave character of electrons. Ultimately, all of these new, unfamiliar ideas were incorporated into a full-blown mathematical theory of atomic phenomena. The new quantum theory correctly explained many atomic phenomena and the periodic table of the chemical elements; and it went on to predict lasers, transistors, and antimatter. The early hypotheses seemed to resist any intuitive interpretation that could reconcile them with the classical conception of matter. In fact, the interpretation of quantum theory became even more puzzling and unintuitive as it evolved to its present sophisticated form.

THE COLOR OF HEAT

If you place a metal poker in a fire and let it heat up for a few minutes, it will have a visible red glow when you remove it. Our observations of heated objects like pokers have prompted us to use terms like *red-hot* and *white-hot* in everyday parlance. In fact, all bodies glow when they're heated (if they don't burn up first). Even objects at or below room temperature radiate electromagnetic waves that are not visible. Only at temperatures above several hundred degrees is the radiant energy of an object visible.

An object that is very dark or black in appearance is an efficient radiator of electromagnetic energy. An ideal or perfect radiator is therefore called a black body.[1] At any given temperature, the character of the electromagnetic energy radiated by a black body is unique. The spectrum or frequency distribution of black-body radiation is uniquely determined by the temperature of the body. For example, a body at the temperature of the sun's surface—about 6,000°C—radiates most of its energy at frequencies corresponding to visible yellow light.[2] Proportionately less energy is radiated at the lower frequencies of red light and also at the higher frequencies of blue light. And even less energy is emitted at the frequencies of invisible infrared and ultraviolet radiation. The unique combination of fre-

---

1. "Black" refers to the fact that a perfect radiator is also a perfect absorber of energy, and thus it appears black because it reflects none of the light that strikes it. When heated sufficiently, however, a black body will radiate visible light and appear red or yellow or white.

2. Just as the frequency of a sound wave corresponds to the pitch we hear, so the frequency of a light wave corresponds to the color we see.

quencies in the spectrum of sunlight appears to our eyes as white light. It is that same white sunlight that Newton's prisms first broke apart into its component colors and then recombined into white light. The white light of the sun is characteristic of any object heated to about 6,000°C.

A cooler black body at a temperature of about 4,000°C will radiate most of its energy at the lower frequencies of red light and thus appear red-hot. The frequency spectrum of the light from a heated body can tell us the temperature of the body. Astronomers routinely analyze the spectra of starlight to determine the surface temperature of stars; and cosmologists, as we saw in Chapter 6, have similarly analyzed the spectrum of the cosmic background radiation to determine its temperature.

This correlation between the temperature of a body and its radiation spectrum is qualitatively understandable from Maxwell's theory of electromagnetic waves. The theory predicts that a charged particle like an electron will radiate electromagnetic energy when it is accelerating or changing speed. In matter, atoms and other charged particles are accelerating because they are constantly vibrating back and forth, going faster and slower. The forces that hold atoms and molecules together are not rigid, but rather they act like springs that give particles a little leeway to vibrate back and forth. Since vibrating charged particles are accelerating, they radiate electromagnetic energy, according to Maxwell. The faster the particles vibrate, the more energy they radiate.

It turns out that the temperature of a body is a measure of the average speed of the vibrating atoms in the body. As the temperature increases, the atoms vibrate more rapidly and radiate more energy, and so we expect a correlation between temperature and radiation. Maxwell's theory, in fact, predicts how much energy is radiated at each frequency. The only trouble is that the prediction is completely wrong.[3]

---

3. Maxwell's theory predicts an infinite amount of energy in the ultraviolet part of the spectrum. This is known as the ultraviolet catastrophe.

PLANCK'S QUANTUM HYPOTHESIS

In 1900, Max Planck correctly calculated the frequency spectrum for black-body radiation by proposing a new model for the vibrating atoms in matter. Planck assumed that atoms in matter could not oscillate or vibrate at all possible frequencies. According to Planck, atomic oscillators could vibrate only at specific allowed or *quantized* frequencies. The term *frequency* refers to the rate of vibration of a wave or of an oscillator, like a vibrating spring. A typical sound wave vibrates a thousand times a second. The spring in a car's shock absorber may vibrate once every two or three seconds. An FM radio signal vibrates about 100 million times per second. Atoms vibrate many times faster than radio signals. Springs, sound waves, and radio signals can all vibrate continuously over a certain range of frequencies. In fact, what we call *sound* is a traveling pressure wave in air that vibrates anywhere between 20 and 20,000 times a second. A normal ear can sense sounds throughout this range. Each frequency we hear corresponds to a different pitch, and the ear is capable of hearing all the pitches continuously over about thirteen musical octaves.

Many musical instruments do not produce all possible pitches within their range. In contrast to a violin, a piano can produce only certain pitches, or frequencies of sound. You can play a B or a B-flat on the piano, but nothing in between. On a violin you can play tones between B and B-flat, although a violinist well trained in Western music will avoid them like the plague. On a piano or a xylophone you have no such option. The B is 494 oscillations per second and the B-flat is 466. You cannot play a note of 471 or 486 oscillations per second. You might say a piano is quantized: it can play only certain allowed pitches or frequencies; other frequencies are not possible. A violinist can produce a continuous frequency spectrum of pitches. A pianist, however, is limited to a discrete frequency spectrum of quantized pitches. Much of Western music is based on the quantized set of twelve specific tones in a musical scale. Only those twelve tones and no others are allowed in musical compositions. In music, we're used to this—in fact, we expect it. But no one ever expected it for atomic oscillators. Planck's assumption of quantized atomic oscillators to explain black-body radiation was unprecedented.

By assuming that the atomic oscillators in a body are quantized, i.e., that they can oscillate only at certain prescribed frequencies, Planck was able to calculate the correct spectrum of radiation from a black body. The vibrating charges in Planck's model also radiate energy as in Maxwell's theory, but Planck's quantized atomic oscillators give rise to a spectrum of radiated colors, or frequencies, that correctly agrees with the observed spectrum of frequencies radiated by a black body. Planck's model clearly worked better than Maxwell's theory. But what was the cause of Planck's assumed quantization? Why were vibrating atoms quantized in the first place?

THE PHOTOELECTRIC EFFECT

What makes a photographic light meter sensitive to the brightness of light? How does an electric eye open a door? In these devices, light enters a photocell, where it is converted to an electrical current that can energize a meter or open a door. This remarkable conversion of light into electricity not only opens doors. In 1905 it also opened the new quantum era of physics.

When light strikes the surface of a metal, it knocks electrons out of the metal atoms. This phenomenon, known as the photoelectric effect, had been carefully observed and was well known empirically before 1905. Like black-body radiation before it, the photoelectric effect could not be explained by Maxwell's theory. The problem for Maxwell was with the colors or frequencies of the light.[4] We find that with certain metals, ultraviolet light is

---

4. Recall that color corresponds to the frequency of light.

capable of ejecting electrons that move very rapidly. Blue light can eject electrons, but they move more slowly. With yellow light, electrons come out that are slower still. And red light cannot expel electrons from the metal at all. So the high-frequency ultraviolet light produces the fastest electrons, while the low-frequency red light cannot provide electrons with enough speed to escape from the metal. Yet Maxwell's theory incorrectly predicts that if the light is bright enough, it should expel electrons from the metal, and that it is brightness, not frequency, that determines the speed of the expelled electrons.

Furthermore, the wave theory of light cannot properly account for the ejection of the electrons. Waves of light spread out from their source like the expanding circular ripples from a pebble dropped into a pond. As the light waves spread, they become more and more diffuse. The tiny portion of a wave that encounters a minuscule electron carries very little energy. It cannot provide the concentrated wallop needed to eject the electron from the metal. So the Maxwell wave theory of light cannot explain how the electrons get enough energy to escape from the metal or why their ejection speed depends on the frequency of the light.

To overcome the failures of Maxwell's theory, Einstein made two assumptions. First, he assumed that the light was concentrated into little packets of energy, called *photons*, which acted like particles rather than waves. Each compact photon of light carries enough energy to collide with an electron and knock it out of the metal. Second, Einstein extended Planck's idea and assumed that the photons themselves were quantized. Each photon has only one specific frequency, and it carries an amount of energy exactly proportional to its frequency, neither more nor less. Ultraviolet photons have twice the energy of the red ones because they have twice the frequency. Thus, ultraviolet photons have enough energy to eject electrons from a metal with plenty of speed to spare. But red photons have too little energy to liberate any electrons. In the intermediate case, if blue photons can expel electrons, they do so only at slower speeds. Furthermore, increasing the brightness of the light increases the *number* of photons but not their energy. Bright red light consists of many low-energy photons, none of which is capable of ejecting electrons. With ultraviolet light, even of very low intensity, there may be few photons, but each photon has more than enough energy to expel an electron.

DUALITY

Einstein was able to explain all the features of the photo-electric effect by assuming that light acted like quantized particles and not like waves. Despite its incomprehensible particle picture of light, Einstein's explanation was so compelling and its influence so far reaching that it won him the 1921 Nobel prize for physics. Light, it seems, has a dual nature—at times acting like waves and at other times like particles.

How was this possible? Waves and particles are contradictory concepts. Electrons, rocks, and planets can be treated like particles; sound and light like waves. Particles are very different from waves. Particles are idealized bodies whose size can be ignored in a given context. The size of the earth is very small on the scale of the Solar System, so it can be treated like a tiny concentrated particle of matter orbiting the sun. The same is true of an electron in an atom. It isn't the absolute size that allows us to treat an object as a particle but its size relative to its environment. An object whose size is negligible in a certain context can be treated like a *point* particle—an idealized concentration of matter at a single point in space. A particle is thus an idealization that has a precisely defined location because it has no extension in space.

By contrast, a wave is diffused over a large region, like water waves on a lake or sound waves in air. It has no well-defined position, nor does it itself consist of matter. The wave is not the medium (material or otherwise) but a movement of the medium. As an ocean wave passes a floating raft, the raft bobs up and down but does not move in the direction of the traveling wave. Only the wave, but not the water, moves toward the shore. A wave is a vibration or disturbance traveling through a medium.

The differences between particles and waves could hardly be more pronounced. Particles are concentrated, localized, and consist of matter. Waves are diffuse, unlocalized, and are not themselves material but vibrations of media. How can light be both a wave and a particle? How can light be both concentrated and diffuse, localized and nonlocalized, material and nonmaterial? And yet Einstein's new analysis of the photoelectric effect demanded a particle picture of light.

Nor is the photoelectric effect unique in treating electromagnetic radiation as particles. High-frequency photons, like x-rays and gamma rays, act just like concentrated little billiard balls when they collide with protons and electrons. Such phenomena that treat light as a particle seem to defy the well-known wave behavior of electromagnetic radiation. Light undergoes diffraction and interference, which are phenomena that uniquely characterize waves. The Maxwell theory, despite some shortcomings, had achieved phenomenal success with a wave picture of light. It seems that light has a dual nature—acting like waves in certain phenomena and like particles in others.

This dual nature of light, and of electromagnetic radiation generally, is a characteristic of all quantum phenomena. Furthermore, as we shall see, duality also characterizes the quantum treatment of electrons, protons, and all subatomic particles. They have a wave nature as well as a particle nature. All matter and energy in the quantum world is dualistic. It combines the conflicting and contradictory properties of both waves and particles.

Yet quantum theory never actually resolves this conflict, nor even acknowledges it *as* a conflict. We begin to see now why quantum theory is so peculiar and unfamiliar. Quantum theory does not actually interpret nature for us in the way that classical physics and relativity do. Newton's physics describes a planet orbiting the sun, and we have a clear visual picture of this process, which corresponds precisely to Newton's mathematical description. Einstein's description of planetary motion is conceptually more abstract, but we can still visualize the process by geometrical analogy, and the mathematics of relativity can always be translated into what we actually see in the heavens.

Quantum theory, by contrast, makes no such concessions to visualization. It correctly predicts the photoelectric effect and the interference of light, but it provides no picture of either. In fact, it claims that there are no "deeper" pictures to begin with. It tells us the results of experiments and observations but not why they come out as they do. In quantum theory we treat nature like a black box. If we press buttons and move levers on the box, quantum theory correctly tells us what the meters on the box will read. But it never tells us what is inside the box, why the meters read as they do, or what goes on inside the box between our readings. According to quantum theory, the box has no inside. No wave or particle picture is needed to predict the results of our experiments with light, and there is no deeper picture of light to be discovered. Whatever conflict there is resides not in the predictions but in their interpretation—in the efforts of human beings to visualize what is going on. Quantum theory denies that phenomena have any inner reality. It provides answers only for the results of actual experimental observations, and it tells us nothing about what happens between our observations. Therefore, quantum theory claims that science can provide no pictures of the inner workings of nature.

ELECTRON WAVES

In the photoelectric effect, we saw that light, which we've always thought of as a wave, acts like a particle or photon. Well, if a wave can act like a particle, then perhaps a particle can also act like a wave. In 1924, Louis de Broglie

considered this possibility. He suggested that an electron—
a particle—could act like a wave. De Broglie's conjecture
turned out to be correct. (He later received the Nobel prize
for it.)

Since 1924, much evidence has accumulated to support
de Broglie's hypothesis. A very good example is the phenomenon of electron diffraction. A beam of electrons
strikes a crystal inside a cathode-ray tube (essentially a
television picture tube). The beam is then diffracted, or
dispersed, as the electrons interact with the regularly
spaced layers of atoms in the crystal. This diffraction
shows up characteristically as a symmetrical circular pattern on the face of the cathode-ray tube. The electron
diffraction pattern is almost identical to the pattern a laser
light beam produces when it is diffracted by the evenly
spaced wires in a mesh screen. The electron pattern is also
similar to the pattern produced by water waves striking a
row of regularly spaced wooden piles at a wharf. Diffraction is a phenomenon unique to waves, and the fact that
electrons can be diffracted is clear evidence of the validity
of de Broglie's hypothesis. Quantum theory incorporates
this dual nature of electrons as a matter of course, but it is
no more comprehensible than the duality of light.

The Spectra of Atomic Light

While the new quantum hypotheses for the microworld
originated in the attempts to understand such phenomena
as black-body radiation, the photoelectric effect, and electron diffraction, its sweeping general power was demonstrated in its ingenious explanation of the atom and all
atomic spectra.

By the second decade of the twentieth century, our
early thinking about the atom had evolved into the planetary model, which views the atom like a miniature solar
system. The negatively charged electrons are imagined as
orbiting the positively charged nucleus. The positive nucleus electrically attracts the negative electrons, just as the

sun gravitationally attracts the planets and keeps them in captive orbits. This model had achieved a modest level of success in explaining the behavior of atoms and matter, but once again it was light that brought down the planetary model.

We saw in the case of black-body radiation that a vibrating electrical charge is accelerating and therefore must radiate electromagnetic energy according to Maxwell's theory. The orbiting electrons in the planetary model of the atom are also accelerating.[5] Therefore, Maxwell's theory predicts that electrons in the atom must radiate electromagnetic energy.

There are two problems with Maxwell's prediction. First, because the electrons are radiating energy away, they are losing energy. So the electrons should slow down and spiral rapidly into the nucleus, causing the atom to collapse. But this patently contradicts the fact that most atoms are quite stable and have been around for billions of years. The second problem with the Maxwell prediction has to do with the light emitted by atoms. Maxwell predicts that as electrons spiral into the nucleus, they radiate light in an unbroken sequence of colors or frequencies—what we call a continuous spectrum. The observed spectrum of atomic light, however, is not continuous, but discrete.

This contrast between a continuous and a discrete spectrum is exactly the same one we encountered when illustrating the idea of quantization with musical instruments. The notes on the piano are quantized into a scale of separate distinct tones—a discrete spectrum of pitches or frequencies. The notes on a violin, by contrast, may be played in a continuous series of linked pitches—a continuous frequency spectrum. Similarly, the white light from the sun or from an incandescent bulb can be separated by a prism into a smooth unbroken band of colors—a continuous frequency spectrum. But the spectrum of light from

---

5. Recall that curved motion is accelerated motion, as was discussed in Chapter 3.

an atom is not continuous; it is discrete. It consists of a series of separate and distinct colors or frequencies. Maxwell's theory, however, was at a complete loss to explain the discrete character of atomic spectra.

THE BOHR ATOM

It was Niels Bohr who first applied the quantum hypothesis to the atom and gave an explanation for the discreteness of atomic spectra. Bohr simply denied the validity of the classical planetary model of the atom with its orbital electrons radiating energy continuously and spiraling into the nucleus. Just as Planck had quantized atomic oscillators in a black body and Einstein had quantized light in the photoelectric effect, so now Bohr assumed that the electrons in an atom were also in a quantized condition or state. By this Bohr meant that electrons could not orbit in any arbitrary path about the nucleus. Instead, only certain specific orbits are allowed, and in each allowed orbit, the electron has a fixed amount of energy—no more and no less. Like the quantized atoms in a black body that vibrate only at certain allowed frequencies, the electrons in Bohr's atom are quantized and have only certain allowed energies. While in these quantized orbits, the electrons are said to be in a stable or "stationary" state. They cannot radiate or emit any of their energy.

There was no precedent for Bohr's quantized states of electrons in the atom. In the classical planetary model, the accelerated electrons had to radiate energy. But in Bohr's quantum atom, electrons do not radiate in their stationary states. Clearly, the electrons in Bohr's model were not little particles revolving around the nucleus. In fact, Bohr immediately recognized the inconsistency between his quantum model and any classical picture of electrons in the atom. Even at this early stage, Bohr was beginning to advocate a view of quantum theory devoid of pictures and interpretation. We may talk of quantized orbits, but this should not lull us into thinking of orbiting particles. It's better to speak of electrons existing in a nonpicturable quantized state (sometimes called an *orbital*), and not to think of electrons as *particles* in that state. Bohr was already anticipating de Broglie's duality hypothesis for electrons.[6]

The radiation we observe from atoms results when an electron makes a transition, or "jump," from one quantized stationary state to another. This is the original quantum jump, but the term is misleading. The electron simply ceases to exist in one state and immediately begins to exist in another. There is no physical movement through space. Once again, quantum theory gives us no picture of the inner workings of the atom. The word *jump* must be interpreted as a nonspatial transition from one quantum state to another.

The energy of each stationary state is different. They are quantized like the pitches, or frequencies, on the piano. In passing from a state of higher energy to one of lower energy, an electron loses energy. The lost energy, in the Bohr model, is emitted or radiated from the atom as a photon of electromagnetic energy or light.

In the hydrogen atom, for example, an electron may jump or drop from an "excited" or high energy state to the lowest energy or "ground" state. If the electron drops from the first excited state down to the ground state, the emitted photon has a specific energy, corresponding to the energy difference between the two states. From Einstein's analysis, we know that the energy of the photon determines its frequency. So the photon emitted from hydrogen when the electron drops from the first excited state to the ground state has a specific frequency. Similarly, when an electron drops from any state to a lower state, a photon with a specific frequency will be emitted. The atomic hydrogen spectrum consists of all the photons emitted in all the

6. De Broglie's wave hypothesis for electrons came some ten years after Bohr's model of the atom. Only later did it become clear that electrons in the atom were acting more like waves than particles. The electron waves, however, are not physically detectable and give us no visual picture of the electron.

possible electron jumps from one state to another. Each of these photons has a unique distinct frequency because any two hydrogen states have a unique distinct energy difference between them. Only these specific photons (having specific colors if they're photons of light) will appear in the spectrum. So the atomic spectrum of hydrogen is discrete and not continuous.

Bohr's atomic model deposed the classical planetary picture, which had incorrectly predicted the collapse of the atom and continuous atomic spectra. But the Bohr model, and the complete mathematical quantum theory that was its legacy, forever rejected any visual or conceptual images for atomic phenomena. Quantum theory provides us with remarkably accurate quantitative predictions of atomic phenomena, but it denies us any picture of the inner workings of nature. Visualization is the price we pay for quantum accuracy. A rather worrisome trade-off.

The Periodic Table of the Chemical Elements

The final triumph of the Bohr model, as refined by the new quantum theory of the 1920s, was the sense it made for the first time of the periodic table of the chemical elements. Chemists had long been aware of certain repetitive patterns among the chemical elements. The halogens— fluorine, chlorine, bromine, and iodine—have similar chemical properties and reaction patterns, as do carbon, silicon, and germanium. The noble gases—helium, neon, argon, krypton, and xenon—are all relatively inert chemically and do not easily react with other substances. Each of these families of elements with common properties falls into one vertical column of the periodic table. The eight columns, or families, in the table thus demonstrate among the chemical elements a regular repeating pattern, which was a complete mystery before the advent of quantum theory. The quantized states of the atom provided a clue to the discrete repetitive structure of the periodic table. The problem was to discover how the quantized states are filled as we add electrons to the atom and how the elements are ordered in the periodic table.

The key was discovered by Wolfgang Pauli. Despite all the quantized states that are available in an atom, the electrons ought to favor the ground state of lowest energy. This is because any physical system always seeks its lowest level of energy. So the six electrons in the carbon atom or the ninety-two electrons in the uranium atom should all crowd together in the ground state, thus giving the carbon or uranium atom its lowest possible energy. Pauli realized, however, that there must be electrons in the higher states in all atoms. Clearly something is operating in the atom— some principle—that prevents all the electrons from crowding into the ground state. Pauli called this the *exclusion principle* because it excludes electrons from any given quantum state once that state's quota is filled.

As we move through the periodic table, each element has one electron more than its predecessor. As the electrons are added, they first fill the ground state and then occupy states of progressively higher and higher energy. A quota is determined for each state by the mathematical rules of quantum theory—two electrons are allowed in the ground state, eight in the next higher state, and so on. As the quota of each state is filled, the electrons are forced into higher states. It's a kind of quantum version of the rule that says two bodies cannot occupy the same place at the same time. The exclusion principle acts like a repulsive force that keeps any more than the allowed number of electrons from occupying any one quantum state.[7]

Following the quantum rules for the states and the Pauli exclusion principle, the order of the elements in the periodic table occurs as if by magic. Chlorine falls right

---

7. Actually, each atomic state contains a certain number of component substates, and the rule is that only one electron is allowed in any given substate.

PERIODIC TABLE OF THE CHEMICAL ELEMENTS
Illustrating the Electron States of the First 11 Atoms

PERIODS

I	II	III	IV	V	VI	VII	VIII
HYDROGEN							HELIUM
LITHIUM	BERYLLIUM	BORON	CARBON	NITROGEN	OXYGEN	FLUORINE	NEON
SODIUM							

into place just below fluorine, and neon below helium. What's more, the quantum states for elements in the same column have a very similar pattern, which in turn explains why they have similar chemical behavior.

For example, lithium is the third element. The ground state in lithium is filled by the two allowed electrons; the third electron must follow the Pauli principle and occupy the next higher state. So lithium has only a single electron in its second, or outermost, state. The next seven elements—beryllium, boron, carbon, nitrogen, oxygen, fluorine, and neon—complete the quota of the second state with eight electrons. These eight elements also complete the second row of the periodic table.

The next element is sodium, which begins a new row and falls into the column under lithium. In sodium, the eleventh electron must move to the next state; so sodium, like lithium, has a single electron in its outermost state. Sodium also appears below lithium in the same column of the table. Chemically, lithium and sodium are alkali metals that react very violently with water and easily combine with the halogens to form salts—lithium fluoride, sodium chloride (table salt), and so on. Lithium and sodium exhibit similar chemical behavior because they both have a single electron in their outermost state—what chemists call a single valence electron.

The behavior of a chemical element is determined for the most part by its valence or outermost electrons. Now we see why lithium and sodium undergo similar or repetitious chemical reactions, and we also see why the repetition occurs and why there is a periodic structure among the elements. The repetitive periodic character of the chemical elements simply reflects the recurring symmetries of the quantum states of the atom.

CHAPTER 9

The New Quantum Theory

DURING THE 1920s, THE LOOSE COLLECTION of new ideas about photons, electrons, and atoms was synthesized into a full-blown mathematical quantum theory. The first great success of the new theory was a complete quantitative description of the simplest atomic system—the hydrogen atom. Quantum theory went on to ever greater achievements, bypassing all the dead ends encountered in Newtonian mechanics. Only quantum theory was capable of handling the details of atomic and molecular phenomena in explaining the large-scale behavior of matter. The new theory, however, raised even more puzzling paradoxes than the wave-particle duality. Instead of Newtonian certainty and determinism, quantum theory answers our questions with probability and statistics. Classical physics told us precisely where Mars was to be found. Quantum theory sends us to the gambling table to locate an electron in an atom. Then there's Heisenberg's uncertainty principle, which places an ultimate absolute limit on our knowledge of the microworld and tells us that we can make no measurement without affecting the result. Bohr gives us his complementarity principle, which raises duality and other contradictory descriptions to the status of a universal law of nature.

NEWTON ECLIPSED

Beginning with the publication of *Principia Mathematica* in 1687, Newton's system of mechanics rapidly achieved a profound and widespread influence on scientific thought.[1] In one grand synthesis, Newton had solved the age-old problem of planetary motion and had created into the bargain a complete mathematical system for describing the general motion of all material bodies. Physics as we know it today was truly born in 1687.

Newton provided a rationale for the motion of Mars in terms of the gravitational attraction of the sun. His description of that motion was accurate enough to predict the course of Mars in the heavens for centuries to come. Using Newtonian mechanics today, we can determine the relative positions of the sun, moon, and earth so precisely that astronomers confidently outfit and embark upon eclipse expeditions without the slightest hesitation or doubt. Weather notwithstanding (alas, we cannot predict the weather by using Newtonian mechanics), even a billion-dollar tourist industry puts its complete faith in eclipse predictions.

But what works for the sun does not work for electrons. The quantum theory description of an electron in a hydrogen atom bears no resemblance whatever to the precise orbits of Newtonian physics. Instead, we are given information about the probability of locating the electron somewhere within the space of the hydrogen atom. This information is in the form of what's called a *probability distribution*, which in turn is based on the quantum-theory wave description of the electron.

---

1. I. Newton, *Mathematical Principles of Natural Science*, translated by A. Motte, revised by F. Cajori (Berkeley: University of California Press, 1962).

THE WAVE FUNCTION OF AN ELECTRON

In Chapter 8, we noted that in his new quantized model of the atom, Bohr had anticipated de Broglie's hypothesis of electron waves. The classical idea of a particle orbiting the nucleus is completely inconsistent with Bohr's nonradiating stationary states of the electron in the atom. The idea of a *standing wave*, however, turns out to be quite useful in describing Bohr's stationary states. The vibrations of a plucked string, like one on a guitar, are called standing, or stationary, waves. Standing waves appear to stand or remain at rest on the string in contrast to traveling waves that move along a string or over the surface of a lake.

What all waves have in common is their characteristic geometrical shape—an undulating curve whose ups and downs represent the crests and troughs of the wave. The mathematical description of the shape of a wave is called a *wave function*. It is a formal mathematical statement that gives an abstract description of the wave as it travels through space (or as it "stands" and vibrates, in the case of a standing wave). A wave function can be represented visually in a graph, which shows the shape and contour of the wave, but the graph is not necessarily an actual picture of the wave. A wave function does represent the actual appearance of a wave on the ocean or on a guitar string, but it does not represent a picture of a sound wave or an electromagnetic wave, neither of which is directly visible.

An electron in an atom is described by a wave function. The electron wave function is analogous to that of a standing wave on a guitar string. It's as if an electron exists somehow in the atom as a vibrating standing wave, and while in this "stationary" state, it does not radiate any energy. Of course, this wave description does not represent the physical appearance of the electron, which is a meaningless idea, according to quantum theory. But the quantum description of electrons as standing waves is consistent with the known fact that electrons do not radiate any electromagnetic energy while in their normal state in the atom. What does it mean, then, to speak of an electron as a vibrating standing wave?

In quantum theory, as we've seen, the very idea of images or visual interpretations is moot. In fact, quantum waves are neither physically detectable, nor are they waves in any kind of physical medium. Sound waves are vibrations in air and can be sensed directly. Light waves are vibrations of an electromagnetic field, and although such a field is not a material medium, its physical presence can be measured directly. But quantum waves are a purely mathematical abstraction. The wave function of an electron does not represent the physical electron at all. Instead, it is a source of information about an electron—a kind of mathematical almanac. By manipulating the wave function mathematically, for example, you can obtain what's called an electron's *probability distribution*, which is a kind of tabulation of how the probability for locating the electron varies or is distributed throughout space. In other words, you find out where the electron is most likely to be located in the atom.

The probability distribution for an electron in the ground state of the hydrogen atom might tell you, for instance, that the electron has a 40 percent chance of being found at a certain distance from the nucleus but only a 10 percent chance of being found twice as far from the nucleus. The probability for locating the electron varies from place to place: it is distributed or proportioned differently at different points in space. If you look for the electron in the 40 percent region, then an average of four times out of ten you'll find it there. If you look in the 10 percent region, then your chances are one in ten of finding it. This is the best information and the only information on electron location that you can ever have, according to quantum theory.

The wave function of the electron (and its corresponding probability distribution) never tells you how an electron behaves or even exactly where it is. Instead of a precise Newtonian orbit, you get gambling odds on where the electron is most likely to be found. Furthermore, the probability distribution gives you not the slightest clue as to what an electron looks like or how it can act as it does. By the same token, you can study the probability distribution for the game of craps from now until doomsday without ever learning what a pair of dice looks like or how it behaves. To know that the chances for snake eyes is 3 percent, for sevens is 17 percent, and for elevens is 6 percent is to know the probability distribution for craps. This same distribution also gives you the odds on a pair of wheels of fortune that point to the numbers one through six. If you know only this distribution, you have no idea what the gambling mechanism looks like. It might be a pair of dice or spinning wheels—and there are other possibilities. Similarly, a wave function is not a picture of an electron.

So, you ask, what then is an electron wave function? It is an abstraction, a mathematical reservoir of information. It provides a formulized means for obtaining information about any physical observation you can make or any experiment you can perform. But like an information table, which you might consult about Congress or Antarctica, it never gives you a picture of its subject. It gives you the readings on the black box but never reveals its inner mechanism. Some physicists even deny the existence, or at least the relevance, of such inner mechanisms. Quantum theory, they claim, tells us all there is to know. No one has ever made a measurement that quantum theory could not correctly predict. Anything else—any conjectures about what electrons look like or their ultimate nature—is no part of science. What quantum theory does give us is primarily statistical and probabilistic information, so quantum theory may be described as a theory of information rather than a theory of nature. This is a radically new conception of science.

THE UNCERTAINTY PRINCIPLE

In an effort to clarify some of the apparent contradictions that quantum theory seems to harbor, Werner Heisenberg

announced in 1927 a new principle of uncertainty or indeterminacy.[2] Heisenberg had been troubled by the wave-
particle duality of an electron. If an electron behaves like a
wave and has no precise trajectory in space, then how can
we see, for instance, the path of an electron in a cloud
chamber? When an electron travels through the supersatu-
rated alcohol vapor of a cloud chamber, it leaves behind it
a vapor trail that marks its passage. Because the electron
moves very rapidly, the vapor droplets form long after it
has passed through the chamber.[3] In effect, the droplets
provide clues about where the electron was, but they do not
mark its precise path. After all, the droplets may seem very
small to us, but they are millions of times larger than the
electron itself. The droplets pile up along the trail to create
a thick smoke screen, as it were, which conceals and blurs
the actual trajectory of the electron.

Although Heisenberg began thinking about an electron track in a cloud chamber, he came to realize that the
problem of locating an electron is not one of instrumentation and is not peculiar to the cloud chamber. No matter
what experimental apparatus Heisenberg considered for
"tracking" an electron, he found it was never possible, as it
is for a moving particle in classical physics, to measure the
precise path of an electron: for example, the droplet size is
too large in a cloud chamber. Or if you try to "see" an
electron with light (or x-rays), then the light photons strike
the electron like billiard balls and randomly change the
electron's position. And so on. Heisenberg found that there
was no way—even with the most perfect experimental
apparatus—to pin down precisely an electron's trajectory.
It was not an experimental problem but a limit *in principle*
on the accuracy of observing an electron. The tiny size of

---

2. An excellent book that presents Heisenberg's own personal views
on the evolution and significance of quantum theory is W. Heisenberg, *Physics and Beyond* (New York: Harper & Row, 1972).

3. The electron will pass through the chamber in a few millionths of
a second, but the vapor droplets will take thousandths of a second to
form.

the electron always prevents you from measuring its position and motion with perfect accuracy.

Heisenberg found that there is always a *theoretical* limit—an irreducible degree of uncertainty or indeterminacy—in measuring the simultaneous position and motion of an electron. The uncertainty principle is the general statement of this canny observation of Heisenberg's. It is impossible to make a precise determination of an electron's path. There will always be some uncertainty when you try to measure both the position and motion of an electron simultaneously. You can never completely eliminate the errors in these measurements. In fact, the two measurements—of position and motion—counterbalance one another. The better you measure the position, the worse you will be able to measure the motion, and vice versa. If you could measure the position (or motion) perfectly, you would know nothing about the motion (or position).

The uncertainty principle thus seems to reconcile the particle-wave duality. The better you know the position of an electron, the more localized it is and the more it seems to act like a particle. Alternatively, if you know the motion or speed very well, then the electron's position is poorly known. The electron has a well-determined speed but is diffuse like a wave. Whether or not this actually resolves duality is questionable. The uncertainty principle offers a different point of view on duality but still leaves it as a paradoxical fact of life. Indeed, the uncertainty principle raises even deeper and more puzzling issues in science.

Why, after all, should there be any ultimate limit on the ability to measure an electron? Is nature hiding its secrets from us, or isn't there any nature there to do the hiding? Make no mistake about it—the uncertainty principle is unique and unprecedented in science. In classical physics, we were well aware of errors in measurement. Even the orbit of Mars is not determined with perfect precision. But these "classical" errors were always experimental. They were caused by the limitations of our equipment. With ingenuity, we could always reduce experimental errors to a tolerable or negligible level. But Heisenberg confronts us with something entirely new in scientific law—an ultimate theoretical limit on knowledge. No improvement in experimental technique, no refining of the methods of observation, no brilliant ingenuity can ever eliminate the uncertainties in measuring the trajectory of an electron (or of any other subatomic particle, for that matter). For the first time in the history of science, a natural law tells us that there is a fundamental limit on what we can know. The uncertainty principle reinforces the idea that quantum theory gives us no picture of the inner workings of nature—no look inside the black box.

The Loss of Objectivity

But Heisenberg isn't finished with us yet. There are even more astonishing implications of the uncertainty principle. Although we can never measure the electron with perfect accuracy, we can determine what we will observe— a wave (if we refine our motion measurement) or a particle (if we refine our position measurement). What we observe depends on how we observe it. The observed phenomenon depends on the observation, and therefore on the observer. Thus quantum theory raises doubts about one of the most deeply held articles of scientific faith—the belief in an objective world that is independent of our observations. Quantum theory and the uncertainty principle tell us that we cannot picture nature but can only predict the results of specific experiments—and these results will depend on how we choose to perform our experiments. Not only are we blind to the workings of nature, but even our brief glimpses are of no objective, independent reality but of a subjective, observer-determined world.

Complementarity

It was Niels Bohr who guided a generation of physicists in the 1920s and 1930s toward an acceptance of the paradoxes of quantum theory. Bohr's philosophy developed gradually over a period of years in consultation and debate with many other physicists who came to study and work at Bohr's Institute for theoretical physics in Copenhagen. Consequently, the philosophy born and nurtured in the throes of these disputes, with Bohr acting as midwife and nurse, came to be known as the Copenhagen interpretation of quantum theory.[4]

One particular series of debates—between Bohr and Einstein—dealt with the reality of the quantum world, as Einstein struggled (unsuccessfully) to find flaws and deficiencies in the basic structure of quantum theory, which he was never able to accept intuitively as a proper theory of physics. Bohr always succeeded in disproving Einstein's ingenious arguments and counterexamples. The Bohr-Einstein debates became celebrated in the annals of modern physics and ultimately led to remarkable refinements in the validation of quantum theory and to some incredible innovations in its interpretation.[5]

According to the Copenhagen interpretation of quantum theory, the duality and nonobjectivity of the microworld were inevitable facts of life. Bohr looked toward a future science that would be radically different from classical physics. From the earliest days of his atomic model, Bohr had realized that there was no way to reconcile our classical pictures of nature with the physics of the microworld. The apparent conflict between the wave picture and the particle picture of the electron, or between position and motion in the uncertainty principle, suggested to Bohr the existence of a deep principle of nature. He called it the principle of *complementarity*. It states that our descriptions of the microworld present mutually exclusive views that are inconsistent with each other but which are complementary.

---

4. A book that attempts to reconstruct a coherent view of Bohr's ideas is D. Murdoch, *Niels Bohr's Philosophy of Physics* (Cambridge: Cambridge University Press, 1989).

5. See M. Sachs, *Einstein versus Bohr* (La Salle, Ill.: Open Court, 1988).

The different views complement each other in the sense that all views are needed to form a complete picture. Like complementary angles or colors, or complementary intervals in music, the whole (a 90° angle, whiteness, or a musical octave) is formed from a combination of its complements, or parts.

The complementary views of light as either a wave or a particle are contradictory only if we attempt to picture what happens to light between our observations of it. In the rainbow dispersion of light by a prism, light acts like a wave, not a particle. In the photoelectric effect, light acts like a particle, not a wave. There is no conflict in any one observation or experiment. The conflict arises only in our efforts to reconcile the results of different experiments. We think of light as something with its own integrity. We believe light is independent of our observations and obeys consistent laws of behavior. According to Bohr, thinking of light as independent represents our classical biases about nature. Such thinking has no place in the microworld, which obeys the laws of quantum theory, not Newtonian mechanics. Quantum theory correctly predicts the results of any experiment we can perform with light or electrons. There are no contradictions. Our conjectures about the "nature" of light or electrons lie outside the realm of science. To help us think about phenomena, which we can do only in classical terms, we must recognize that neither the particle picture nor the wave picture provides a complete view of light. They complement each other, and both are required for a complete conception of light.

SCHIZOPHRENIC ELECTRONS

An even more dramatic illustration of complementarity is the electron two-hole experiment. A beam of electrons in a cathode-ray tube is fired at a shield plate that contains two small, closely spaced holes through which the electrons can pass. The electrons that emerge from the holes produce a visible pattern on the face of the cathode-ray tube. The pattern does not show two spots as we might expect. Instead, it looks similar to the pattern we discussed in the electron diffraction experiment in Chapter 8. In this case, it's called an interference pattern and refers to the effect of two waves combining or interfering with each other. The interference of light waves is responsible for the rainbow reflections we often see on oil slicks on a wet road; the crisscrossing wave pattern we can observe from time to time on the surface of a lake is due to the interference of water waves. This interference pattern on the tube face indicates that the electrons are acting like waves. We get essentially the same pattern when a light beam passes through two holes.

But an electron can't pass through two holes at the same time, you may say in protest. So we do another experiment to discover through which hole each electron passes. We place detectors at each hole that signal every time an electron passes through. But when we know through which hole the electrons pass, the interference pattern disappears! The pattern is now that of individual electrons passing through single holes without any interference. What's more, if we "thin out" the electron beam so that *only one electron at a time* passes through the apparatus, then we still see an interference pattern in the first case and an individual pattern in the second case. A single electron acts like a wave and interferes with itself when both holes are open, or else it acts like a particle when we detect it at the holes. An electron does indeed seem able to pass through two holes at once. (At least, that's the interpretation we are forced to make if we insist, contrary to quantum theory, on describing the electron when we're not observing it.)

In the first version of the experiment, the electrons act like waves and produce an interference pattern. In the second version, we localize the electrons, they act like particles, and the wave interference pattern disappears. The two experiments complement each other. Neither gives us a complete description. The full complement of electron behavior is provided only by the results of all electron experiments. The results of any one experiment are never contradictory, because we never see an electron acting like a wave and a particle *at the same time.*

CAUSALITY COMPLEMENTED

The most radical implication of complementarity is what it tells us about causality and space-time. The information that quantum theory gives us is statistical and probabilistic. We cannot locate an electron exactly. We can know only the relative likelihood of finding it in different places—40 percent here, 10 percent there, and so on. This suggests that electrons do not obey causal laws as planets do in Newtonian physics. The position of Mars in space is precisely determined at each instant of time by Newton's laws. Such classical behavior exemplifies causality—specific effects are inevitably determined by certain causes. The causal deterministic laws of Newtonian physics give us precise information about the space-time behavior of material bodies. Causality operates within the framework of space-time, as we saw in Chapter 5.

In quantum theory we have no causal behavior of electrons in space-time. The quantum description of electrons in space-time is statistical. An actuarial table can tell us with great precision how many Americans of a certain age will die next year, but it cannot identify a single American who will die. Actuarial information is statistical. It can predict trends and averages with great accuracy, but it cannot provide specific causal information about any individual. Neither can quantum theory. If you ask for a space-time description of an electron, you get statistics, not causality. Does that mean that the laws of quantum theory are not causal? No, it does not. The equations of quantum theory are causal to the core. But it is the causality of wave functions and not of electrons. Quantum theory describes the causal evolution or development of wave functions, which are mathematical abstractions, as we have seen. Wave functions are neither physical nor detectable, but they are causal.

Actuarial trends also obey causal laws—the mathematical laws of statistics. But actuarial tables are not people and do not die. Actuarial laws are causal. Deaths of people are statistical. In quantum theory, causality and space-time descriptions are complementary. You can have one or the other—wave function causality or space-time statistics, but not both at the same time. The classical idea of a phenomenon unfolding causally in space-time is outlawed in quantum theory. Space-time and causality are complementary descriptions.

MACRO VS. MICRO

How is it possible, then, that Mars does follow deterministic causal laws? Hasn't quantum theory replaced Newtonian physics? And why don't we have a wave function for Mars? As a matter of fact we do. But the wave function for Mars cuts no ice. Mars is much too big.

The de Broglie wave hypothesis and the uncertainty principle relate the quantum effects of a body to its mass. The larger and more massive a body is, the less pronounced is its wave nature. The tiny electron has a wave function of substantial size on the microscale. Electron waves can be diffracted by objects comparable to them in size, like the atoms in a crystal. But large bodies have wave functions of incredibly small size. The quantum waves associated with Mars or a baseball—or even a speck of dust, for that matter—are so minuscule in size that no physical object is small enough to diffract them. Even a quark is trillions of times too large to diffract a Martian wave. In effect, the quantum nature of macroscopic objects is undetectable, and they act, to all intents and purposes, like classical particles that obey causal laws. Quantum theory, in principle, does apply to galaxies as well as to electrons. But quantum effects for objects larger than molecules is so negligible that they are indistinguishable from the predictions of classical physics.

If Mars, boulders, people, and even specks of dust are too big for quantum theory, and if electrons, atoms, and molecules are squarely in the quantum domain, then where is the boundary between the quantum and classical worlds? There is no simple answer to this question. Quantum effects can be magnified up to the everyday domain as is attested to by semiconductors, microchips, and superconductivity. Many commonplace properties of matter on the large scale, like electrical conductivity and transparency to light, can be explained only with quantum theory, as we shall see in Chapter 14. But as far as individual objects are concerned, one must simply examine each individual case to see whether quantum effects are important. In general, it will depend not only on the size of the object but on the circumstances as well. No sharp universal boundary can be drawn. For the foreseeable future, we shall continue to send people to the moon using the laws of Newtonian physics and to probe the secrets of matter with quantum theory.

We knew nothing about quantum theory until the twentieth century because we had no significant experience until then of the microworld of atoms and electrons. The paradoxical weirdness of quantum theory has no effect on our bodies. But it still plays havoc with our minds.

CHAPTER 10

Metaphors of Matter

AS WE EXPLORE TWENTIETH-CENTURY physics, the nature of space and matter becomes more and more problematic. Physicists of the eighteenth and nineteenth centuries tried to explain all natural phenomena in terms of the motion of matter in space. Stars moved within galaxies, planets within the Solar System, atoms within objects, and, ultimately, electrons within atoms. Substance or matter was composed of tiny impenetrable particles that "occupied" space and moved within it. Space itself, separate and distinct from matter, was empty, fixed, and absolute—the bare stage on which the material particles performed the dance of the phenomena. But Einstein forever blurred the sharp distinction between space and matter. In relativity, space is fluid and variable and its very geometry is equivalent to matter. In quantum theory, electrons are described through wave functions, which are not vibrations of matter but of a mathematical medium of "information"—a complete abstraction. The former notion of matter as substance has become pure stuff and nonsense. Our changing ideas of space and matter in physics represent a profound evolution of metaphors.

THE LITERAL AND THE METAPHORICAL

The idea of metaphor, as used in this and earlier chapters, may seem out of place in a book on physics. In its simplest sense, *metaphor* refers to a figurative comparison that evokes an ironic, humorous, symbolic, or poetic similarity between two different and unlike things. "Life is a dream" and "crocodile tears" are examples. As poets and artists know only too well, metaphor is the very stuff of creativity. We can hardly think without analogies, comparisons, and metaphors, and metaphors are far more common in science than we realize. Notions of the universe or the human body as a machine and of the mind as a computer have become part of the vernacular. But even such basic physical concepts as space, time, matter, and number have an essential metaphorical character.[1]

To speak of the curvature of space as a gravitational field is unquestionably a metaphor. Some physicists may protest that this comparison represents a precise mathematical equivalence. Curved space and gravity are one and the same in the equations of relativity. Metaphor is a linguistic concept that has no place in mathematics. But if space and gravity are one and the same, then why must we equate them? Equality implies an equivalence between two different things; a relationship between two different things is exactly what a metaphor is. Mathematical equations effectively function as metaphors. If we try to sever semantics from mathematics, then we can never interpret an equation or a mathematical theory. Language is as essential to the practice and progress of physics as is mathematics.

A sharp distinction between the literal and metaphorical, even in science, is impossible to draw. One scientist's

---

1. See R. S. Jones, *Physics as Metaphor* (Minneapolis: University of Minnesota Press, 1990).

meat is another's poison. Even if we could divest mathematical equations of all metaphorical content, it would still be impossible to discuss, report on, and—most important—create science without the use of metaphors. All creative work employs metaphor. The fruitful blurring of the literal and metaphorical that characterized medieval thought has largely been denied and rejected in the modern scientific age.[2] But the curved space-time of relativity and the complementarity of quantum theory have returned metaphor to a central role in the poetry of twentieth-century science.

MATTER AS STUFF

The idea of tiny impenetrable entities, or particles, as the fundamental constituents of matter dates back to the ancient Greeks in the time of Democritus. Our very word *atom* comes from a Greek term meaning undivided. At the dawn of the modern scientific era, it was Descartes who made the strongest claim for the motion of particles as the fundamental basis of all phenomena. Although Newton rejected many of the details of Descartes' work, he retained its basic spirit of a mechanics of particles. By the end of the nineteenth century, all material phenomena were assumed to result from the motion of particles and of electromagnetic waves in space.

Although atoms, electrons, and later neutrons and protons were taken to be fundamental at one time or another, the history of twentieth-century physics has been the history of an ever deeper penetration into matter and the discovery of ever smaller particles as the elementary constituents of matter. The "indivisible" atom was split long ago into its electrons and nucleus. The electron has miraculously survived unscathed from the penetrating rays of ever more powerful atom smashers. It remains indivisible and elemental in most current theories of physics. The

---

2. For an illuminating analysis of the contrast between medieval and modern metaphorical thought, see O. Barfield, *Saving the Appearances* (London: Faber & Faber, 1957), especially Chapters 11-14.

nucleus has not been so fortunate. It was broken apart into neutrons, protons, and other exotic members of the so-called *hadron* family. Neutrons and protons are not themselves fundamental particles. They, in turn, are made of much smaller particles called *quarks*—the latest and somewhat elusive candidates for the most elementary subnuclear particles.[3]

Despite all this dividing and conquering, we haven't yet the slightest idea what elementary particles are made of. Electrons and quarks are the smallest and most fundamental objects we can detect (quarks only indirectly). Quarks are a thousand times smaller than the neutrons and protons they inhabit and one hundred million times smaller than the once indivisible atom. What this means is that down to the smallest scale we can measure with our particle accelerators, electrons and quarks remain indivisible and without any internal structure. If electrons and quarks are made out of anything else, it's still too small for us to "see."

Down to the smallest scale of current-day measurement, electrons and quarks are point particles—that is, they have no size. Now the idea of a point particle—a tiny bit of matter that has no size or extension in space—has long been a useful idealization in physics, a metaphor. As a physical reality, it's an impossibility. How can matter occupy no space at all? To be truly an impenetrable point object, matter would have to have an infinite density—a physical impossibility. And if some tiny bit of matter could occupy no space, then why couldn't a larger chunk of matter—or the whole universe, for that matter—do the same? Why would matter take up any space at all? Sooner or later, quarks and electrons will have to reveal some size

---

3. Electrons and the five other members of its lepton family, together with the family of quarks, are taken to be the fundamental constituents in the currently fashionable "standard model" of matter. But the hypothetical superstring, if it should ever prove to be real, suggests a yet deeper and more basic level of matter. More on all of this in Chapter 18.

and structure if we are to avoid an explanation of the nature of things in terms of a physical impossibility. As a metaphorical ideal, the point particle has been a great boon to physics; as literal reality, it is a profound embarrassment.

Even if we can get around the size problem, what is the stuff that elementary particles (whatever they may be) are made of? Is there anything of substance down there? As we have seen, the quantum-theory description of an electron is in terms of a wave function, which is neither material nor even physical. Quantum theory gives us some information (which is only statistical, remember) about the location of an electron, but it tells us nothing about what it is we find (if we're lucky enough) and what the damn thing is made of. On these subjects quantum theory is mute. Its only concession is the laconic and maddening comment that we have no need of such information.

Well, if wave functions aren't matter, they are at least mathematics. Is everything then "made out of" mathematics? What does that mean? And is mathematics preferable to metaphor as the flesh and blood of the universe? Mathematics, like metaphor, is an artifice of the human mind. We marvel and rejoice over the mathematical structure of the universe. Everything from the cosmos to the quark seems to manifest a mathematical structure or pattern. But if the universe is "made of" mathematics, then what other structure could it have? And if it is made of mathematics, is the universe then a product of the human mind? And is the universe a metaphor for . . . what?

Perhaps we go too far. (Or is it physics that goes too far?)

This Too Too Solid Flesh

If we can't figure out what the stuff is that matter is made of, can we at least get some insight into its solidity and impenetrability? The old adage says you can't put two things in the same place at the same time. The quantum-theory version of this, as we've seen, is the Pauli exclusion principle, which says that two electrons cannot both occupy the same quantum state in an atom. So both the solidity and incompressibility of matter and the periodic structure of the chemical elements depend on the idea that atoms or molecules or electrons are impenetrable. But if matter or electrons aren't made of some substance or stuff, then what is it that cannot be penetrated?

We need to be careful. The impenetrability of matter isn't just a matter of stuff. It's also related to rigidity—to what keeps matter from collapsing and makes it occupy a certain amount of space. There are forces between atoms that hold them in place and keep them from getting too close. These attractive and repulsive forces, as they're called, are electromagnetic in nature. Their combined action in condensed matter keeps atoms at a certain average distance from one another.[4] Thus the shape, rigidity, and extension of objects is explained in terms of interatomic and intermolecular forces.

There is a logical problem here, however. In our discussion of gravity in relativity, we noted that the force of gravity is never directly observed. We infer its existence from the motion of planets, projectiles, and falling rocks. The hypothesis of a universal force of gravity enables us to explain many different phenomena under one rubric—from the behavior of galaxies to Archimedes' principle for floating bodies. We never actually see the force of gravity—it is known "only by its fruits." This fact, together with the impossibility of distinguishing gravity from acceleration, was Einstein's key to the general theory of relativity, which dispensed with the force of gravity altogether.

Now matters are essentially no different for the electromagnetic force (not to mention the other basic forces of nature that operate only inside the nucleus of the atom). Forces are observed and measured in terms of their effects.

---

4. The term *condensed matter* refers to matter in the solid or liquid state, in contrast to the much more diffuse gaseous state.

When we see a compass needle being deflected, we attribute the needle's rotation to the magnetic force of attraction of the earth. We never see or directly observe this magnetic force, only its effect. All our information about the character and structure of the earth's magnetic field is inferential. Similarly, we attribute the rigidity, solidity, and impenetrability of matter to electromagnetic forces, which, of course, are never directly observed. In science, this is not a problem, because the assumed existence of electromagnetic forces explains so many different phenomena that we hardly think of it as an assumption. The theory of electromagnetic forces hangs together as a self-sustaining system of great generality and utility.

The Logic of the Matter

From the logical point of view, however, there is a problem. To claim that electromagnetic forces explain the rigidity of matter and yet to infer the existence of these forces *from* the rigidity of matter is a logical fallacy—a circular argument. Science often gets away with such circular argument. We can infer electromagnetic forces from the observation of colliding electrons, say, and then use the same forces to "explain" solidity. But there is a big stretch here— using one phenomenon to explain another—especially when the second phenomenon involves billions of atoms in an interaction far more complex than the simple repulsion of two electrons.

Furthermore, the forces assumed to explain the rigidity of matter are always associated with matter itself, whose form and substance, as we've already seen, are complete mysteries. Forces and matter go together in physics. We never have one without the other. To explain the rigidity of matter in terms of its internal forces is like saying that matter is solid because it is solid. The forces are not directly observable and can never be dissociated from matter itself. Forces are intrinsic properties of matter. Their existence and character is no more or less of a puzzle than that of matter itself.

The impermanence of forces is clearest in the theory of relativity. There, geometry replaced gravity. All the inferences from planets and falling bodies could not save the force of gravity, once Einstein had made a better and more encompassing theory (also, of course, an assumption). Had Einstein succeeded in the great quest of his final years to geometrize electromagnetism, as he had done with gravity, we would no longer talk of an electromagnetic force. But, of course, he failed. Quantum theory made its reputation on the basis of its treatment of electromagnetism. Quantum electrodynamics is touted as the most accurate theory in history. And so for the moment, the electromagnetic force is secure (if illogical).

In the final analysis, even if forces could be placed on a firm logical, as well as empirical, basis, they could not save the day. For interatomic and intermolecular forces clearly do not explain electrons and quarks. If these elementary particles are not points (which is a physical impossibility), then what holds them together or keeps them from collapsing, as the case may be? Why doesn't the electrical charge of an electron repel itself and fly apart? If we assume another force to hold it together, then we're back to the same circular argument.[5] Forces are no better at explaining solidity and extension than the abstract descriptions of matter itself.

The whole problem was neatly summed up many years ago by Bertrand Russell:

> The matter that we construct is impenetrable as a result of definition: the matter in a place is all the events that are there, and consequently no other

---

5. Such paradoxes and dilemmas are central to the controversy over the procedure of "renormalization" in quantum electrodynamics and particle physics. See, for example, H. R. Pagels, *The Cosmic Code* (New York: Simon and Schuster, 1982), 292.

event or piece of matter can be there. This is a tautology, not a physical fact; one might as well argue that London is impenetrable because nobody can live in it except one of its inhabitants.[6]

THE HEART OF THE MATTER

Physics presents us with a bewildering array of theories, models, and metaphors for matter: Newtonian matter in space, Einsteinian matter as space, gravity as acceleration, gravity as geometry, particles as waves, waves as particles, wave functions, probability distributions—and there are other examples that we'll encounter in later chapters. The changing view of matter, first as substance, then as space, and finally as mathematical abstraction or information, demonstrates a profound evolution of metaphors. Our concept of matter becomes more and more ethereal. The objects of the everyday world are evaporating before our very eyes (our mind's eyes, at least). We can hardly distinguish in physics any longer between space and matter.

This strange dematerialization of matter into space, geometry, and information is not unique to science. In Platonic philosophy, the cosmos itself was an organic whole—a living being. Its space was like a cosmic realm of intelligence, consciousness, and wisdom. There was no sharp distinction between the cyclic motions of the heavenly bodies and the thinking processes of the mind. Motion in space, changes in time, and thinking in the mind were related and intertwined in ways we can no longer comprehend.[7] All of this represents a metaphorical state of mind that was still present in medieval astrology but which we no longer experience as a culture. We think of such a consciousness as primitive, superstitious, anthropomorphic. But aren't we begging the question?

In Hinduism, there is also the concept of the *akasha*,

---

6. B. Russell, *The Analysis of Matter* (New York: Dover, 1954), 385.
7. See Barfield, *Saving the Appearances*, Chapter 15.

which is ether or space and yet is a kind of vast warehouse or library of all past, present, and future events. The *akasha* incorporates space and time but also the life force itself and, most significantly, the medium of hearing and of revealed knowledge.[8] Here is a remarkable blend of space, matter, and consciousness, which despite its ancient primitive origins seems to hint at both the mathematical informational representation of matter in quantum theory and the spatialization of matter in relativity. Hindu philosophy sees the material world as illusion—and how far is illusion from abstraction and mathematics? What, after all, are our twentieth-century castles in the air? Mind, mathematics, metaphor . . . ?

---

8. See Daniélou, *The Gods of India* (New York: Inner Traditions, Ltd., 1985), 49.

CHAPTER 11

Interpretations of Quantum Theory

FROM THE OUTSET, THE UNIQUE and revolutionary character of quantum theory provoked a vigorous debate over its validity and interpretation, which has continued throughout the twentieth century. Are the laws of nature ultimately probabilistic? Is there any reality there between our observations? Do we actually affect phenomena by observing them? Einstein was foremost in leading a crusade on several fronts to discredit quantum theory. For one thing, he refused to accept laws of nature that were no different from gambling odds. He could not believe that "God plays dice with the world." In a celebrated series of debates with Bohr, Einstein raised another group of criticisms in which he attempted to expose potential loopholes, fallacies, and deficiencies in quantum theory. Bohr ingeniously managed to rebuff all of Einstein's attacks. But in a 1935 paper, Einstein marshalled a new argument that claimed that quantum theory fails to provide a complete description of nature. According to Einstein, certain physical effects can be observed that quantum theory fails to predict. This challenge ultimately led to a remarkable series of experiments by Alain Aspect that were designed to settle the dispute. Aspect's experiments thoroughly vindicated quantum theory. The theory does in fact predict, but cannot explain, some very weird phenomena, which Einstein had assumed were impossible. It seems that information can travel faster than the speed of light—apparently in direct violation of relativity and the laws of causality. The implications of Aspect's experiments remain controversial, but they have prompted even more outlandish interpretations of quantum theory, which no doubt would have amused and exasperated Einstein, had he lived to hear about them.

COPENHAGEN REVISITED

The Copenhagen view of quantum theory, developed primarily by Bohr and Heisenberg, has generally come to prevail among the majority of physicists, despite continuing controversy. According to this view, the laws of nature are neither objective nor deterministic. They do not describe a reality that is independent of the observer. You cannot help but affect the outcome of any observation you make, as the uncertainty principle tells us. The experimental predictions that the laws do make are always statistical rather than deterministic. No trajectories, only probability distributions.

The seeming contradiction between electron waves in one experiment and particles in another is an example of the complementarity principle. Quantum theory predicts wave or particle behavior correctly and consistently, but never simultaneously. According to Bohr, the contradictions arise only in our minds when we try to infer an ongoing "behind-the-scenes" picture of an electron, which is no part of the theory. Furthermore, there is nothing else to be learned from nature other than the limited and statistical information that quantum theory provides. The theory is complete: anything it does not tell us may be interesting conjecture or metaphysics, but it is neither observable nor measurable, and therefore is irrelevant to science.

Nothing about the Copenhagen interpretation satisfied Einstein's requirements for what a properly objective and deterministic theory of physics ought to be like. Over the years, he challenged and debated Bohr through a series of thought experiments designed to demonstrate some inconsistency or error in the predictions of quantum theory. Einstein would present Bohr with some dilemma or contradiction in quantum theory. Bohr would sleep on it and return the next day with a resolution. Invariably, Einstein had overlooked something or neglected some effect. In one case, ironically, Einstein forgot to take into account his own general theory of relativity.[1] Ultimately, Einstein accepted the internal consistency of quantum theory, but he clung tenaciously to one crucial criticism.

It was Bohr's claim that the theory is complete that seemed most vulnerable to Einstein. He could not accept the idea that electrons weren't real things that exist in physical reality, whether we observe them or not. If quantum theory cannot describe continuous elements of physical reality, then clearly the theory must be incomplete. In 1935, Einstein and two colleagues wrote a paper that challenged the completeness of quantum theory.

THE EINSTEIN-PODOLSKY-ROSEN ARGUMENT[2]

The paper by Einstein, Podolsky, and Rosen has become so well known among physicists and philosophers of science that it is usually referred to simply as EPR. The idea of the EPR paper is to offer a thought experiment that could be done in principle but whose outcome cannot be predicted by quantum theory. If there is an experimental measurement that can be made but that quantum theory fails to

---

1. See M. Sachs, *Einstein versus Bohr* (La Salle, Ill.: Open Court, 1988), Chapter 7. A more detailed account of the debate can be found in P. A. Schilpp, ed., *Albert Einstein: Philosopher-Scientist* (New York: Tudor, 1949).
2. A. Einstein, B. Podolsky, and N. Rosen, "Can Quantum Mechanical Description of Reality Be Considered Complete?" in *Physical Review*, vol. 47 (1935): 777.

predict, then the theory is incomplete. A complete theory must be able to predict the result of any actual experiment.

I shall present a simplified version of the original argument, which is nevertheless faithful to the spirit of EPR. We imagine two electrons colliding like billiard balls. Now as we know, electrons are not literally particles, but they can act like particles under certain circumstances. Electrons cannot actually touch and bounce like balls with hard surfaces, but because electrons carry negative electrical charge, they repel each other. Two electrons that approach each other head on will repel and thus slow each other down more and more as they get closer. Finally, the electrons will reverse their motion and fly off from each other in opposite directions. Even though they never touch, they act as if they had bounced apart.

In both classical physics and quantum theory, a collision like this between two bodies is governed by the law of conservation of momentum. Applied to this case, a conservation law states that some physical quantity, like momentum or energy, is conserved or unchanged in the course of the collision. The direction and the speed of each electron may change, but the total momentum of the two electrons will remain constant. The total momentum is simply the sum of the momenta of the two electrons. The momentum of each electron is a certain measure of its motion,[3] so as the two electrons "collide," the law of conservation of momentum governs their interaction. Their total momentum before and after the collision must be the same.

Particles that interact in this way according to the law of conservation of momentum are said to be correlated. The momentum of one electron correlates with the momentum of the other. If, before the collision, one electron has 2 units of momentum and the other has 5 units of momentum, then their total momentum before the collision is 7 units. By conservation of momentum, the total

3. The momentum of a particle is defined as the product of its mass and velocity. It is a characteristic measure of motion. In particular, it is the measure used in Heisenberg's uncertainty principle.

momentum after the collision must also be 7 units. If, after the collision, the first electron now has 3 units of momentum, then the other must have 4 units. By knowing the correlated total momentum before and the momentum of one electron after, we can find the momentum of the second electron after the collision. This very powerful technique can be used to solve such problems as firing a ballistic missile and tracing the temperature of a gas to the interaction of its molecules. We shall use it now to understand the EPR argument.

The two electrons in our EPR thought experiment are correlated by conservation of momentum. After the two electrons interact, they fly apart, but their total momentum remains what it was before the interaction. It turns out that the relative position of the two particles is also correlated. Thus, if I measure the position of electron one, I can infer the position of electron two. The two electrons are correlated by both their momenta and their positions. Suppose after the interaction I measure the position of electron one and shortly afterwards the momentum of electron one. I can perform these measurements with arbitrarily high accuracy because I'm measuring only the position in the first case and only the motion in the second case. (Recall that the uncertainty principle prohibits us from simultaneously measuring the position and motion—momentum, to be precise—of one particle with arbitrarily high accuracy.) In our thought experiment, let's assume the measurements are exact.

Now, here's the rub. The measurement of the exact position and momentum of electron one (at different times) tells me the corresponding exact position and momentum of electron two. (The method is similar to the calculation of correlated momenta that we did above.) But I have never disturbed electron two, and therefore I could not have affected its position or momentum in any way. Its momentum (or simply, its speed) remains constant as its position varies uniformly. So the position and momentum I have determined for electron two must be its exact position and momentum at one particular time after the collision. Therefore, I now know the simultaneous position and momentum of electron two. This contradicts the uncertainty principle. In other words, I have made a measurement—of the position and momentum of an electron—that quantum theory says is impossible and about which it makes no prediction. Therefore, quantum theory is incomplete. This is the EPR argument in a nutshell.

Einstein Contra Bohr

At first hearing, the EPR argument sounds devastating. Not only is quantum theory incomplete but it's the uncertainty principle—the very cornerstone of the quantum theory's foundation—that is at fault. It *is* possible to measure simultaneously the position and momentum of a particle (except perhaps in Copenhagen).

But EPR never fazed Bohr, for the deeper implications of EPR are completely consistent with Bohr's point of view. Bohr says to Einstein:

"So, Albert, you claim you can measure the exact position and momentum of an electron. Let's see if you are right. Let's try to measure directly both the position and momentum of electron one in your thought experiment. If we do this, the uncertainty principle will surely kick in and prevent us from making the exact measurements you claim. Don't you agree?"

Einstein responds with a twinkle in his eye. "Of course I agree, Niels, if I measure the position and momentum of electron two *directly*. But I can measure the momentum of electron two *indirectly* by using the correlation of the two momenta."

"But, my dear Albert, *have* you indeed measured the momentum of electron two indirectly? Once you measure the exact position of two, its momentum has no meaning, according to quantum theory. You have determined a number, I grant you, but is it really the momentum of two? The only way to check it is by a direct measurement, which you have already admitted you cannot do."

"Surely, Niels, you cannot discount my indirect measurement. I have made two physical measurements of electron one and then used correlations to determine the position and momentum of electron two. This is a perfectly valid procedure, used all the time by experimental physicists. Clearly electron two has a position and momentum, whether I observe it directly or not."

"That is precisely where we differ, Albert. Quantum theory tells us nothing about continuously existing elements of physical reality. It tells us only the results of specific measurements. And you cannot measure the exact position and momentum of electron two, or of any other electron, for that matter. The unobserved 'measurement' of an electron is a figment of your imagination."

"Absurd! Do you mean to tell me that my measurement of electron one implies nothing physical about electron two? Or, worse yet, are you implying that by observing one, I am affecting two? Does my measurement of one somehow change the momentum of two and make it uncertain? This would be an impossibility—a violation of causality, an instantaneous transfer of information, a case of . . . of . . . of . . . of spooky actions at a distance. It's completely preposterous!"

"Now, now, Albert, please calm down. We see these matters very differently. You speak of violations of causality, of—what was it—'spooky actions at a distance.' All of this assumes there is an objective world there—two electrons interacting according to conservation laws, even when we're not observing them."

"But, but . . ."

"Yes, yes, Albert, I know you disagree, but let me go on. Quantum theory tells us nothing of such an objective independent reality. It tells only what our instruments read. You cannot break the system down into electron one, electron two, and the instruments of our observation. In quantum theory these form an inseparable whole. The instruments and how we use them are a part of the system. We have no right to speak of what electron one or two may be doing if we are not directly observing them. And when we make an observation, the experimental apparatus is automatically included as part of the system. We cannot separate the parts and say how one part is affecting another. All we can do is make a measurement at any given time and ask quantum theory for the result of that measurement, which it unfailingly gives us. But it gives us the result for the totality of the electrons and the experiment—a totality that is whole and indivisible. When you measure the position and then the momentum of electron one, that is all you know. Measuring the position and momentum of electron two is a different experiment with a different outcome. The two results are complementary and incompatible. In the first experiment, I do not know the physical state of electron two, period. I cannot speak of one part affecting another, nor can I try to correlate the results of the two different experiments. There is no violation of causality because no deterministic causal laws operate within the system. Quantum theory does not describe continuous causal behavior of electrons in space-time."

Einstein, now subdued, sits for a moment nodding his head and then responds:

"What you are saying, Niels, is that we have here a deep incompatibility—and it's not only between the two of us. If quantum theory is complete, then there can be no objective causal reality. According to you, my argument demonstrates not the incompleteness of quantum theory but rather its strange complementary character. My picture of the correlated electrons interacting causally in space-time directly contradicts the assumption that quantum theory is complete. Objective causal reality and quantum completeness are incompatible, or, as you would say, complementary."

"Precisely. Space-time and causality are complementary descriptions. The laws of quantum theory do not apply to an objective independent reality. Your EPR argument simply affirms all of this. It demonstrates the validity of the Copenhagen interpretation."

"But are you satisfied with such a state of affairs? No physically real electrons? No causal space-time behavior? No longer a theory of nature, but merely of information?"

"What can I say, Albert? I did not make the laws. Quantum theory answers any experimental question we can ask. It has correctly predicted or explained atomic spectra, lasers, antimatter, semiconductors, superconductivity, and who knows—perhaps in the future the big bang itself. I can hardly ask for more."

"But I must. I must seek a theory that is aesthetically satisfying, too. And it will never be quantum theory. Perhaps I ask for too much, but I seem to have no choice."

We must leave Albert and Niels with their conflict unresolved and go on to see how the physics community at large has dealt with these issues.

A Theorem and an Experiment

In an effort to clarify the confusion surrounding the EPR paper, a British physicist, John Bell, developed a clear mathematical statement of the conflict discussed above by Einstein and Bohr. Bell was able to demonstrate the logical incompatibility between the predictions of quantum theory and any assumed causal behavior among elementary particles. This is often referred to as Bell's theorem. Based on his theorem, Bell was able to suggest specific experiments that would distinguish unequivocally between quantum probability and space-time causality. For practical reasons, Bell modified the original experiment suggested by EPR, which was too difficult to do in practice. But the new experiments would demonstrate even more dramatically the same conflict between inferential measurements and the uncertainty principle.

A series of experiments based on Bell's suggestion were carried out in Paris in the 1980s by Alain Aspect and his associates. Rather than measuring the momentum of electrons, Aspect measured the polarization of photons, which can be done much more accurately. Polarization is subject to a different conservation law—the conservation of angular momentum, which governs the spin and rotational or angular motion of material bodies. The basic idea was the same—to observe the polarization of two correlated photons (rather than the momentum of two correlated electrons), and to see whether measuring one photon had any statistical effect on the other.

Aspect's results as interpreted by Bell's theorem vindicated the probabilistic predictions of quantum theory, beyond any reasonable doubt. The experiments seemed to show that one photon somehow "knows" what the other is doing. Measuring the polarization of one of the photons has an apparent effect on the other, even though there is no mechanism by which they can "communicate." Uncertainty or probability seems to pass magically from one photon to the other and to affect its statistical behavior. In one of the experiments, there was not enough time for a signal—even at the speed of light—to travel from one photon to the other. It seemed impossible for the photons to "communicate" with and influence each other. Yet they did. Einstein's "spooky actions at a distance" seem to be real and to occur instantaneously.

EPR had always assumed that if a physical quantity (momentum or polarization, for example) can be measured without disturbing the system, then that quantity must have an independent objective reality. But neither the inferred momentum of an electron nor the inferred polarization of a photon can exhibit such an independent objective existence. EPR had assumed objectivity and causality and concluded that quantum theory was incomplete. Bell and Aspect, however, had demonstrated the completeness of quantum theory. One must therefore choose to give up either objectivity or causality (or both). Not much of a choice for Einstein (or for the rest of us)!

QUANTUM THEORY A LA MODE

The perfect track record of quantum theory is hard to argue with, and yet the Copenhagen view of nature sticks in the craw of many physicists. Many physicists pay lip service to Bohr but secretly sympathize with Einstein. Few scientists can genuinely accept the idea that they are creating their reality as they observe it. There's got to be something there behind all our experiments and observations; we can't just be making it up. Yet many of the abstractions of quantum theory have the quality of fantasy and imagination. There's no arguing with the magical products of quantum theory or the results of such crucial "philosophical" experiments as those of Aspect. Quantum weirdness and illogic are here to stay. And so a number of imaginative and philosophically-minded physicists have proposed new interpretations in an effort to tame the wild quantum beast.

Heinz Pagels and Henry Stapp claim that there is no real violation of causality.[4] In the Aspect experiments, the polarization of many paired photons is measured. The data for each separate photon is random. We see no pattern or correlations until we compare the data for the pairs of photons. If you examine a mixed deck of playing cards, you will find no special order or pattern. If, however, you should find a second deck with its cards in exactly the same order as the first, you would know that someone had carefully arranged the two decks. But you would never suspect any "arranging" until you examined the two decks together. We can never make use of the "communication" between the two photons. Their mutual "influences" are apparent only long after the fact, when we correlate or compare the data from the two photons. Thus no violation of causality occurs in real time, and no faster-than-light effects can ever be observed directly or used to send infor-

---

4. H. Pagels, *The Cosmic Code* (New York: Simon and Schuster, 1982), 174–76.

mation. Pagels doesn't deny the violation of causality, he simply relegates it to the realm of the unobservable, along with electrons in the atom and motion in space-time. An Einstein, however, might still worry about those spooky skeletons in the closet.

Of course, for Bohr, all of this is a tempest in a teapot. He always insisted that the components of a quantum system, including the measuring instruments, cannot be taken apart or treated separately. The parts cannot "influence" each other. Quantum theory deals only with an indivisible unified whole. It correctly tells us the instrument readings for any given experiment, including Aspect's. There is no use seeking the "mechanism" behind the readings. It has nothing to do with science. Bohr's view is consistent and functional, but a bit Spartan for most people.

Abner Shimony, a philosophically minded contemporary physicist, has a different way of interpreting quantum reality. He argues that probability and chance are not mathematical abstractions but objective properties of nature.[5] The indefinite location of an electron in an atom is not a matter of limited information but of physical reality. The electron exists in a state that is *physically* indefinite. It's as if the odds on a horse race were not a calculation in someone's mind or in a computer but actually existed somewhere in physical reality—on the track, perhaps, or in the horses' hooves!

This differs from Bohr's view. Bohr says we have only statistical information on the whereabouts of an electron. When we look for an electron, quantum theory gives us the odds of finding it. But Shimony says that the electron actually has a statistical character in space-time, which we directly observe. It isn't the information that is statistical, but the electron itself. This contrasts with Bohr, who says we know nothing about the space-time character of the electron.

---

5. A. Shimony, "The Reality of the Quantum World," *Scientific American* (January 1988): 46.

And then there's the many-worlds interpretation of Hugh Everett.[6] There is no contradiction among the various complementary views, nor is there any statistical character to reality. All possible locations of an electron, all correlations of photons, all possible wave and particle manifestations of matter and energy—all of these are occurring together in parallel space-times. The world splits into infinitely many worlds each time an experiment is performed. The many worlds represent the myriad possible results of an experiment that are predicted by quantum theory. The electron goes through the first hole in one world and the second hole in another. There is no probability. Everything that is possible simply happens. Reality is a superspace of infinitely-many constantly dividing worlds. Quantum theory does not give us the odds in a game of craps but rather a travel brochure on the thirty-six "parallel worlds" that split apart with each throw of the dice.

One wonders which is worse: the malady or the cure?

QUANTUM THEORY VS. CLASSICAL PHYSICS

The superior ability of quantum theory to deal accurately and consistently with atoms, lasers, x-rays, superconductivity, and countless other things has allowed the new theory almost completely to eclipse the older classical one. But we still use Newtonian mechanics on the everyday terrestrial scale and even in the space program. The clash persists between the old familiar concepts and the new revolutionary ones.

The laws of the macroworld remain stubbornly deterministic, while those of the microworld are ironically probabilistic. Our kinematics of projectiles and comets has a pronounced visual character, while atomic descriptions have none. The world of shoes and ships and sealing wax seems to endure between our observations, while electrons and atoms have no actual or physical status beyond our

6. See Pagels, *The Cosmic Code*, 178.

measurements. The interpretations of quantum theory that we've discussed don't alleviate the conflict; if anything, they widen the gap between the two worlds.

For most physicists, these dilemmas present no problem. They go right on about their business, largely ignoring the philosophical debates and existential conflicts. After all, physics works: it accurately predicts natural phenomena and permits us to control them. Philosophy is irrelevant.

But the anguished protests of such great physicists as Einstein and Erwin Schrödinger belie the complacency of the majority. (Erwin Schrödinger was one of the founding fathers of quantum theory, but he came to have serious disagreement with the Cophenhagen interpretation.) The violence that quantum theory does to our deep beliefs and assumptions cannot be sustained. The philosophical and spiritual implications of physics weigh heavily on the human heart. We cannot forever endure an impossible boundary between two worlds.

CHAPTER 12

Objectivity, Measurement, and Reality

QUANTUM THEORY MAY OR MAY NOT describe the inner workings of a "real" microworld, but it certainly makes very accurate predictions about actual experimental observations. Accordingly, most twentieth-century physicists still claim to be dealing with an objective reality. This traditional claim rests largely on quantitative techniques that make possible the consistency, reproducibility, and verifiability of measurements. Because different observers can agree, say, on the length of a table, we infer that the table has an "objective" status, independent of who measures it or how. It seems to exist in physical reality, independent of us and our observations. But the concept of measurement is not entirely free of subjective judgment or of logical fallacies, as we shall see. How then can it be a guarantee of objectivity?

Claims of scientific objectivity are also based on the generality of scientific law. Because the same laws apply to electrons, jet planes, and stars, we assume that these laws embody a universal truth that is independent of observers and lawmakers. But criteria like generality, consistency, accuracy, economy, and simplicity are neither objective nor inherent in nature. They are aesthetic choices, which reveal the subjective judgments and predilections that are inevitably built into science.

THE FOLLY OF THE QUANTITATIVE

The sciences generally, and physics in particular, pride themselves on their quantitative character. The presumption is that quantitative descriptions of things are somehow more meaningful and objective than qualitative ones. This amounts to a bias, which is reflected in our culture at large. Less quantitative fields, such as sociology and literature, imitate physics by trying to be as quantitative as possible. Wars and catastrophes are assessed largely in terms of numbers of deaths and quantity of destruction. Even our primary emphasis on money as a measure of wealth and value is an example of the attempt to reduce everything to quantity. These are presumably the most "objective" measures.

All of this amounts to a kind of folly of the quantitative that influences us in many subtle but powerful ways. The organization of our high schools, colleges, and professional lives confronts us with a difficult choice at a fairly early stage in our schooling. We must decide whether to emphasize the sciences or the humanities in our studies. Most of us naturally gravitate toward one or the other. Some lean more heavily on their logical, empirical, quantitative skills, while others depend more on their intuitive, imaginative, qualitative talents. There is, of course, no sharp boundary between the two approaches, nor do we ever use one entirely to the exclusion of the other. People depend and rely on both kinds of abilities. Nevertheless, by predilection, social indoctrination, and schooling, we tend to choose between the sciences and the humanities in our educational and professional lives.

On the face of it, there seems to be nothing wrong with this. Variety is the spice of life, and the success of our society depends on people with a diversity of interests and talents. The problem, however, does not lie in the existence of these differences but rather in our attitudes toward them. We do not treat scientific and humanistic professions as equals. What is it that our society tends to value, support, and reward the most? It is all the technological wonders and conveniences of modern life—heart transplants, CAT scans, computers, VCRs, lasers, the space program, the modern automobile, and jet travel. Of course, we value architecture, theater, and painting, too, but not to the same degree as technology. This is reflected, for example, in the difference between the annual budgets of the National Science Foundation and the National Endowment for the Humanities, or in the contrast between the typical salary of an engineer and a dancer. There are four Nobel prizes in the sciences—physics, chemistry, medicine, and economics, one for literature, none for art or music.

This exaggerated emphasis on the quantitative has a dehumanizing effect on all of us. It tends to devalue and discredit many subjective experiences that are emotional, psychological, aesthetic, ethical, qualitative, ecstatic, symbolic, philosophical, spiritual, eloquent, imaginative, lofty, passionate, enlightening, or sublime. Only what is objective matters. The dominance of quantity also contributes to the fears and anxieties associated with subjects like physics and mathematics.

But is the reliance on quantity as a guarantor of objectivity justified, even in science?

QUANTITY AND OBJECTIVITY

In physics, the quantitative approach reigns supreme. Physics is the quantitative science par excellence—the most "exact" and thus the most objective of the sciences. Qualitative as well as quantitative descriptions of organisms, rock strata, and compounds may be permissible in biology, geology, and chemistry. But in physics, all natural laws and all objective descriptions of phenomena are stated in mathematical terms, as are the definitions of all physical quantities. The very word *quantity* is used in physics to refer almost exclusively to any relevant or important characteristic of the physical world. Temperature may be defined either empirically—as the length of the mercury column in a thermometer—or theoretically—as a measure of the average kinetic energy of molecules. And, of course, length and kinetic energy are themselves defined quantitatively. In fact, the formal definition of any quantity in physics is not an explanation but a prescription for its measurement. Speed is the distance an object travels divided by the time it takes. From the point of view of physics, the definition of speed is an operational procedure, which specifies the use of metersticks and clocks for measuring distance and time.

Such "operational definitions" are the stock-in-trade of physicists. A physical quantity is a useful objective description only when it can be measured. Measuring something implies assigning a number to it, and numbers are what quantification is all about. Once some property of a body or of matter or of space-time is measured and quantified, it can be incorporated into a mathematical statement—a quantitative law of nature—and then used to make quantitative tests and predictions in the real world. So the ability to measure and the techniques for assigning numbers to things are central to the methods and applications of physics.

Because the laws of physics are expressed almost exclusively in terms of measured quantities, scientists believe that physics describes an objective physical reality, which is independent of our observations and experiments. We have seen that quantum theory raises some serious doubts about this objectivity. But the remarkable track record of physics in dealing with the material world is sufficient proof for most scientists that they are dealing with an objective independent reality. Whether we're dealing with an eclipse of the sun, the properties of a semiconductor, or the range and power of a laser beam, it is the agreement between quantitative predictions and experimental measurements that gives physics its power and reputation.

Such consistently accurate agreement is presumably the result of the objectivity of physics—of the precise correspondence between physical law and the external objective world. If I measure the frequency of an electromagnetic wave, and if I do it carefully enough, then I know that any other observer will independently measure the same frequency. I naturally assume that we are each observing some aspect of an external independent reality. Anyone observing the same wave will measure the same frequency because that wave is not affected by or dependent on the observer in any way. Quantitative measurement seems to guarantee that the wave is part of an independent objective reality.

THE MEASURE OF ALL THINGS

But is measurement truly a guarantor of objectivity? To find out, we are going to examine in some detail the process of making a measurement. We'll take a simple example—measuring the length of a table. We choose the measurement of length for two reasons. First, length is simple to understand, both conceptually and in practice. Second, the measurement of length is basic to just about every other quantity in physics. We know, for example, that speed is defined in terms of the measurement of length and time. All concepts in physics—force, energy, momentum, temperature, pressure—are defined ultimately in terms of a few fundamental quantities—primarily length and time.[1] For convenience, the standard system of measuring units in science takes as fundamental a half dozen quantities and defines everything else in terms of them. But in principle, one fundamental quantity is all we need. The quantitative operational definition of all the quantities

---

1. Even time can be defined *operationally* in terms of a length on a scale or dial, for example.

used in physics can be based on a single fundamental measurement—length. Length is fundamental to all measurement in physics and to its quantitative character. But if the procedure for measuring length is *not* purely objective, then that would repudiate any objective basis for physics.

So let's measure a length, which from the formal operational point of view is equivalent to defining length. There are three steps in the process of measuring length: 1) defining limits, 2) comparing with a standard of length (a meterstick), and 3) counting. What we shall discover is that in each of the three steps, one of two assumptions will be necessary—either a subjective judgment must be made or else we must assume a prior definition of length. Any use of subjective judgment will clearly negate the purely objective character of the process. A definition that is based on itself is circular and therefore a logical fallacy. Furthermore, since a circular definition assumes we know what length is when we're trying to define it, we are incorporating an element of tacit or prior human knowledge into the definition, which makes it subjective. In either of the two cases, subjectivity is unavoidable. Thus, a definition of length that is formally equivalent to the measurement of length can never be objective. And such a definition of length (which is equivalent to the process of measuring length) cannot be used as an objective basis for science.

Step 1: I shall illustrate the three steps by performing a "thought measurement" of the length of a rectangular table—let's say of its longer dimension. First, I must determine the limits (or corners) of the long edge of the table so I will know just what to measure. For example, I need to know exactly where to place the end of my ruler. How can I specify this limiting point? I could simply gauge it by eye, but of course, that would be a subjective judgment. I could use a magnifying glass or a blown-up photograph, but ultimately I'd still have to judge whether the end of the ruler and the table line up. To remove myself from the decision-making process, I might try to specify the exact position of the corner for some sensing device and computer to locate. But to specify a position, I need to give its distances from some points of reference. These distances, however, are themselves lengths, and so I must use the concept of length in defining length. (Not kosher!)

Alternatively, I might try to specify the end of the table in terms of its atomic constituents. But even if I (or some device) could see these atoms, they would be in constant motion, and, in any event, establishing their position or distribution would once again require a prior knowledge of length. I would need to use either human judgment or a measurement of length to decide where the atoms "thin out" enough to constitute the end of the table. There is no way to locate the corners of the table without making a value judgment or else using a previous definition of length. Step 1 is either subjective or circular. (And a circular definition, as we've argued above, is also subjective.)

Step 2: The second step is to compare the specified length of the table to some standard unit of length—meters or inches or light-years. I'll choose a meterstick. Let's forget about the problems of step 1 and assume I have managed somehow to line up the zero end of my meterstick with one corner of the table. In order to compare the length of the table with standard meters, I must now determine the mark on my meterstick that most closely lines up with the other corner of the table. Let's suppose the end point at the other corner of the table (however I manage to determine it) falls somewhere between the 3.25 and 3.26 marks on my meterstick. Either the end point falls very close to one of the two markers or between them. In order to compare with meters and ultimately count them, I must judge the relative position and proximity of the two markers on the meterstick and the end point of the table. For the sake of argument, I'll call the two marks on the meterstick points A and B and the point at the corner of the table P. I must decide whether P is between A and B and then whether P is closer to A or to B.

I could, of course, make this judgment subjectively by eyeballing the three points. But that would mean my comparison is not purely objective. What other choices do I have? To determine the order and relative proximities of points A, P, and B, I must do two things: first, identify and distinguish the relevant points from other points, and second, measure the relative distances between them. Identifying points, or anything else for that matter, is something that human beings do very well, but of course by using subjective judgment. To eliminate the subjective element, I might try to use a computer or other automated sensing device. How do you teach a computer to recognize and identify a point, or any object, let's say an octopus or a particular human face? This is an extremely difficult task. Computers are no match for humans when it comes to pattern recognition. To the extent that it can be done at all, features and characteristics must be specified by their positions—a tentacle here, a nostril there. But specifying position requires a prior knowledge of length, as we've seen, which makes our definition circular again.

To make matters worse, we are dealing here not with recognizing octopi and people but with recognizing mere points. The computer must identify points A and B on a meterstick and P at the end of the table. Points have no characteristics other than their positions, and positions are specified by distances or lengths. Again we are trapped in a circle that requires a knowledge of length in order to define it. Simply identifying points for the purpose of comparing a length to standard meters requires either subjective judgment or a prior knowledge of length.

Once the points are identified, judging their proximity (or relative distances) involves the same dilemma—a choice between subjectivity and circularity (also ultimately subjective). So much for step 2.

Step 3: The third and final step involves counting. How many meters long is our table? Surely, there can be nothing subjective about counting. Make no mistake about it—counting or assigning numbers to things is at the very heart of quantification and scientific objectivity. We base our belief in an objective world on our ability to assign numbers to things unambiguously, consistently, and reproducibly, i.e., to count. But can we actually count without making subjective judgments?

Defining numbers for the purpose of counting is no easy task. The history of mathematics itself is tantamount to the gradual evolution and refinement of the concept of number, which has grown more and more abstract over the centuries. Numbers are best defined today in terms of the theory of *sets*. In mathematics, a set is an abstract generalization of the idea of groups or collections of things—any things at all, in any quantity. A set contains elements, like geese in a gaggle or sand grains on a beach or marks on a meterstick. In trying to size up or measure the elements in a set, we become involved in the definition of number.

Assigning numbers to elements in set theory requires that I make a one-to-one comparison between the members of my set—the geese or the marks—and the elements of some standard set. A standard set of three elements has its "threeness" in common with any other set of three elements—three geese, three marks, etc. This may sound redundant, but I can avoid counting or relying on any previously memorized sequence of integers by comparing the members of each set one by one. Suppose my standard "3" set contains the elements labeled X, Y, and Z. I make a comparison in which I associate X with goose Gloria, Y with Sebastian, and Z with Ferdinand. I find that each and every element of my standard "3" set corresponds uniquely to one and only one goose in my comparison set. Therefore, I know that the two sets have the same number of elements and that I have three geese.

In this way, I can establish the sequence of numbers by using my standard sets. I order them by observing which sets have "leftover" elements in my one-to-one comparisons. The great nineteenth-century German mathematician Georg Cantor used this same procedure to develop an arithmetic of infinitely large sets.[2] You never need a number, only a one-to-one comparison to establish the equivalence of sets.

I can identify and label the standard sets by beginning with the so-called *empty set* (the set with no elements) and adding one element to it at a time. (If all of this seems unnecessarily labored and finicky, I can assure you that defining numbers is one of the most difficult of all tasks in mathematics. It's necessary because we are seeking an objective definition of numbers and counting. And we are by no means doing it fully rigorously.)

So my final task in measuring the length of the table is to count the meters. I must compare the individual meters laid along the edge of the table with the standard sets of 2, 3, 4 (and so on) elements.[3] But before I can make the necessary one-to-one matching, I must first identify uniquely and unequivocally the members of each set— meters and standard elements. The problem of objectively identifying objects—such set elements as geese, octopi, meters, points—is the same obstacle that blocked me in step 2. Either I rely on my intuitive, subjective ability to articulate and identify objects, or I must use a prior definition of length in trying to instruct a computer how to do the same thing. Even if computers could do well at recognizing patterns, they couldn't do so without using length. So even the final crucial step of counting is either subjective or circular. We can't even count objectively.

Thus at each of the three steps involved in measuring length, which is equivalent to the formal operational definition of length, it is necessary in practice to make a subjective value judgment because a circular definition also involves subjective judgment. There is no purely ob-

---

2. See E. Kasner, and J. R. Newman, *Mathematics and the Imagination* (New York: Simon and Schuster, 1956), Chapter 2.
3. My standard sets might be the following, for example: a "1" set with one marble, a "2" set with two pencils, a "3" set with three whales, and so on.

jective way *in practice* to define or perform a measurement of length, or of any other physical quantity.

SUREFIRE PREDICTION

Pure measurement in itself is not objective, which means that science is not objective. We might end the whole matter there, except that science still puts on a good show of objectivity. Scientists can still believe they are dealing with an objective world because of the predictive power and generality of the laws of physics. We can predict an eclipse of the sun or the behavior of superconductors with impressive accuracy. Of course, accuracy in prediction is based on the ability to measure, which as we've seen is fundamentally subjective.

But still, something about the predictive power and generality of scientific laws lures us into believing in an objective world. Such a believer might argue along the following lines: "We may quibble philosophically over the definition of measured quantities in physics, but the remarkable consistency and accuracy of prediction in physics could result only from studying and describing a single objective independent reality. Any other subjective explanation—a mentally dependent or personally invented world—strains reason and requires too many accidental coincidences and arbitrary assumptions. If Mars is a figment of your imagination," the believer in objectivity continues, "then why is it the same as my Mars? In fact, why do I have any Mars at all? And why should two imaginary Marses always appear at the same place in the skies and obey the same law of gravity? Physics works as well as it does simply because it tells it like it is. It describes the real world out there as accurately and faithfully as possible."

But the story of prediction is not quite so simple. Like measurement, it harbors an essential and irreducible subjectivity.

When I predict some future position of Mars for you to

verify, I rely on your knowledge of a number of tacit and conventional assumptions. For example, you must use the same system for measuring, describing, and surveying space as I do. It's no good my giving you the "coordinates" of Mars for a certain future date and hour unless we are completely agreed on how to measure space and time. Is this purely a matter of convention? After all, the centimeters and light-years and seconds are not built into the fabric of space-time. We impose them. We choose the method and basic units of measurement, and this cannot be done purely objectively, as we've already seen. But the element of choice goes beyond the method of measurement. For example, when I use such concepts (or quantities) as distance, velocity, and acceleration in making my predictions, I use quantities that have been chosen for their convenience, reproducibility, and consistency—characteristics that I desire (perhaps unconsciously) to incorporate into my predictive scientific theory.

My predictions (and the whole Newtonian system of mechanics) are based on a mathematical technique for transforming information about forces into information about motion. Newton's laws relate the force of gravity on Mars to the acceleration of Mars. The mathematics of calculus enables us to transform acceleration first into speed and then into position (or vice versa). This reversible transformation from acceleration to speed and then to position is central to the whole method of Newtonian physics. It is based on well-known mathematical procedures (also invented by Newton). But it also depends on the definitions of acceleration, speed, and position.

Why is speed—defined as the distance a body travels divided by the time it takes—chosen as the relevant measure of motion? Why not distance multiplied by time or distance squared divided by time? Other measures of motion can be defined that are useful in certain applications.[4]

---

4. In relativity, the definition of speed as distance divided by time is not so useful as in Newtonian physics, and a different definition is often used.

But the standard definition of speed is the one that works in the transformation of acceleration to speed to position, which is central to Newtonian prediction. In other words, the conventional definitions are tailored to make prediction possible. My prediction of Mars requires that we agree not only on space-time measurement but also on the definitions and descriptions of motion. The whole system of measurement and prediction is designed and engineered to fulfill the goal of prediction.

It's like designing a game. It's no surprise to anyone that checkmate is the usual outcome of a game of chess. The board, the moves of the chessmen, and the rules of the game are all designed with the aim of winning by checking the king. But if one chess player were to ask seriously for a hotel on Park Place, as if he were playing Monopoly, he'd be laughed out of the game (or committed to an asylum). Chess wasn't designed to do everything. Neither was physics.

Physics *was* designed to make predictions, and in this respect relativity and quantum theory are no different from Newtonian physics. If a theory, together with its defined quantities and methodology, cannot make verifiable predictions, then it is not science as it is conceived today—certainly it is not physics.

To argue on the contrary (as a believer in objectivity might do) that physics is not contrived but simply describes what is—physical reality—is to ignore all the choices, conventions, definitions, assumptions, beliefs, and tacit knowledge built into science, all of which could have been selected in countless ways, other than the traditional ways of science. Even the so-called mathematical structure of the physical world—the fact that nature imitates equations and geometry—is something of a hype. What we define today as the physical world is also a matter of choice, and the choices were made largely by thinkers such as Descartes and Galileo. The physical reality we take for granted is primarily defined by successful physical theories—theories whose domain of application and methodology have been created and refined until they fit together like hand in glove. Of course it all works. It was designed to.

The system of science is by now so complex, intricate, and familiar that it is difficult to realize that it was the creation of human minds, involving arbitrary and subjective choices and limits. An analogy might help: If you read an airline schedule, you can find out when you will leave Denver and arrive in Cleveland. If the planes fly on time, the schedule is a kind of predictive system. But it can't tell you when you will arrive at your sister-in-law's house in Shaker Heights. If you complain to the airline, they'll tell you that air travel is limited to designated locations. The schedule wasn't made to solve all problems. Neither was science. But we often forget it.

Science and the questions it chooses to answer are all part of one self-consistent and self-defined system. It is a remarkably useful and valuable system, but its use and value cannot guarantee its truth content or its correspondence to any objective reality. We cannot base any objective or realistic status of science on its ability to make verifiable predictions. That is what it was designed to do, just as chess was designed to lead to checkmate. We cannot argue that the success of science establishes its validity. That it is successful is undeniable, but the ability to accomplish a goal is no guarantee of truth or objectivity. If a baseball team wins the world series, we say that it is a great baseball team, not that it establishes the validity of baseball.

Science is a human activity, and like all human activities, it was created with specific purposes and goals in mind. The purposes and goals were not written down in advance any more than were those of baseball. Science and its goals evolved and became better and better attuned to each other in the course of time. But we can hardly use the finely tuned correspondence between science and its goals as a measure of its validity. Predictive power has subjectivity written all over it. It does not prove the objectivity of science or the reality of the physical world it describes.

GENERATING GENERALITIES

There remains another last-ditch argument that the believer in objectivity might advance: "Even if measurement and prediction cannot guarantee truth, then surely the universality of physical law must. We use the very same laws to describe a falling apple and the orbiting moon. Mars, Jupiter, and ballistic missiles all obey the law of gravity. Isn't that too much of a coincidence to attribute to our subjective choices, designs, and goals for science?"

Here we see the temptation to attribute an objective status to our observations of patterns and coincidences. Generality and universality are not built into nature. We value them, and we favor both the phenomena and theories that accentuate them. The ancient Greeks saw a uniform pattern in the motion of the stars across the skies. For them, it reflected the universal harmony of the cosmos. There were a few exceptions to the general pattern. The disobedient stars were called planets or wanderers by the Greeks. Planets strayed from the beaten path and seemed to violate the universal cosmic harmony. So great was the faith in this universal harmony, however, that Greek philosophers and astronomers strove through the centuries to "save the appearances" of the heavens—to reconcile the apparent irregular motions of the planets with what must be the universal behavior of all heavenly bodies.

The quest for a universal description of the heavens continued for two thousand years until Newton discovered (created?) the universal laws of planetary motion. Newton went the Greeks one better—he applied the same laws on earth and in the heavens. The universality of the Newtonian system is one of its most impressive features. Relativity and quantum theory have eclipsed Newtonian mechanics not only because they provide more accurate predictions but also because they are *more* universal in scope.

But do missiles, planets, and stars really move according to the same laws? Do even the planets? The orbit of Mercury precesses more than that of the other planets because of its closeness to the sun. Each planet is perturbed differently by the size and proximity of its nearest neighbors. Other correction factors require that we *combine* observation, calculation, and theory to achieve the accuracy required for the space program. Theory alone is inadequate. In practice, we cannot apply one simple general law to describe the detailed motion of the planets.

Physicists may argue that the laws are perfectly general but that the number and size of the "corrections" varies from case to case. But once we agree that each case must be treated individually, who is to say what is general and what is particular? To argue that the underlying laws are general and the corrections are particular is to beg the question. We call psychotics abnormal because it suits us to believe that the "normal" majority is not psychotic (despite the fact that social behavior on the large scale is demonstrably psychotic). We could equally well have a theory of human behavior that treats each individual as a unique case that cannot be generalized. Such a theory might be cumbersome and inconvenient, but that would neither prove nor disprove its validity. Indeed, universality is a utilitarian choice as well as an aesthetic one.

As for the missiles and the stars, their motion is affected by so many and such different factors that one could easily raise questions about the whole notion of Newtonian generality.[5] The new science of chaos also suggests that for any but the simplest Newtonian problems, the behavior of objects is so sensitive to initial conditions and to prevailing conditions that the universal laws can no longer make meaningful predictions in practice.

What we really mean by universality is an ordered

---

5. For missiles, we must take into account air resistance, composition, and currents; the earth's rotation latitude and terrain; the shape, density, and stability of the missile—to name only a few. For the stars, we must also worry about the gravitational and electromagnetic characteristics of their local neighborhood, the motion of their parent galaxy and star cluster, and so on. In fact, it is highly unlikely that the detailed motion of most stars can be predicted with any great accuracy.

repeated pattern that departs from some background of disorder or chaos. In universal order it is the unexpected departure from randomness that attracts and delights us, and which we *choose* to favor in our theories. The properties of the chemical elements are not random. They display a repeating order that reflects the universal symmetry of the quantum states of the atom. Of course, no two elements are exactly alike. We notice and emphasize the similarities in developing a universal theory for them. Nor are the quantum states exactly symmetrical. The many electrons in an atom perturb each others' states as do the planets in the Solar System. Only through the use of simplified generalizing assumptions can we arrive at approximate descriptions of complex atoms. Even the mighty computer is stymied in the effort to generalize the complex uranium atom from the simple atom of hydrogen. We naturally opt for commonality, analogy, generality.

Jules-Henri Poincaré, the great nineteenth-century French philosopher and mathematician, says that when we follow our instincts for order, harmony, and universality, we select our facts from the realms of the macroscopic and the microscopic.[6] It is much easier to find the exceptional, the improbable, and the general in what is remote. Newton's generality revealed the unseen order in earthbound motion by studying the remote planets. What could be more unlikely, more universal, than a common law to describe the collisions of electrons, missiles, and stars? Scientists follow this same pattern when they simplify a problem, reduce it to its basic elements, and analyze it with an imitative theory from another area. The very processes of simplification and reduction are guided by human values—by an instinct for unity, simplicity, and generality. These are useful, meaningful, and desirable characteristics in a theory, but they represent creative choices of the human imagination, not proofs of objectivity and reality.

Whether we examine measurement, prediction, or

6. H. Poincaré, *The Foundations of Science* (New York: Science Press, 1913).

generality, we find the stamp of human subjectivity indelibly imprinted on the methods and laws of physics. It may be convenient in dealing with a certain realm of human experience to assume the existence of an objective science and an external independent reality. But convenience and aesthetic choices, no matter how effective and traditional they are, must not be taken for absolute truth. The criteria for judging the value and validity of science cannot be dictated by science itself, but must come instead from a broader region of human understanding and wisdom.

CHAPTER 13

Physics and Consciousness

IN CLASSICAL PHYSICS THERE WAS no need to deal with consciousness. An observer could always minimize or remove all of his or her traces from the observation. There was an objective world out there that was unaffected by human observation of it. At the same time, it was assumed that consciousness, like life itself, somehow evolved out of inert matter. Consciousness would ultimately be explained as the result of complex chemical and physical phenomena in the brain. Because consciousness did not affect phenomena and was ultimately a kind of by-product of the laws of matter, it was quite irrelevant to classical physics.

In twentieth-century physics, however, the mind seems once again to have reared its ugly head. More prominence is given to the role of the observer. In relativity, this is mostly a matter of language. It is the frame of reference, and not its occupants, that determines the properties of space-time. In quantum theory, the observer seems to affect what is observed. Does this mean that consciousness plays an essential role in physical phenomena? Is quantum theory really "user friendly" to consciousness?

The Role of Mind in Newtonian Science

As a result of Newton's brilliant synthesis, a new mechanical picture of the world came to prevail in the eighteenth and nineteenth centuries. The universe consisted of infinite empty absolute space, stretching in all directions. Dispersed through this space were infinitely many particles of matter, whose motions were governed passively by the law of inertia and actively by the law of gravity.

The "active" force of gravity had always troubled Newton. He wanted a purely mechanical or corpuscular explanation for the force of gravity—more like the approach of Descartes. For Newton, the mysterious force of attraction between all particles smacked too much of Aristotelian occultism—the seeking out by bodies of their "natural" place in the hierarchy of the elements.[1] But the great success of Newton's method carried the day. Most scientists (including Newton) believed that the metaphysical problem of the origin of gravity would eventually be solved.[2] For the moment, however, all phenomena were to be explained in terms of the particles and the forces between them.

Such a conception of the universe left little room for the intercession or influence of any form of mind or spirit, whether divine or human. Possibly, God was needed to set the particles in motion, as in Descartes' universe of circulating vortices. But once the initial impetus had been given in the creation, neither God nor anyone else was needed to keep the clockwork universe running and developing its myriad forms.

Thus we have a clear conception of an objective reality, independent of thought or consciousness. Human beings might observe the workings of the universe, even provide a brilliant description and explanation of them. But human influence on those workings was incidental,

---

1. See, for example, T. S. Kuhn, *The Copernican Revolution* (Cambridge: Harvard University Press, 1985), 252–65.

2. The quantum explanation of forces will be discussed in a later chapter. Whether or not it would have satisfied Newton's desire to eliminate occult explanations, however, is another matter.

minimal, and decidedly not fundamental. Consciousness enabled us to study and interpret nature but was not central to it. Indeed, consciousness and the life forms from which it evolved were themselves products of the activity of the inert particles. Consciousness was a secondary, derivative phenomenon in the universe. Only the particles and their interactions were fundamental.

The notion that complex phenomena can be derived from or reduced to simpler fundamental ones is called reductionism. Consciousness can be reduced to biology, biology to chemistry, chemistry to physics, and physics to the activity of the fundamental inert particles. For example, if one had sufficiently detailed knowledge of the behavior of elementary particles, one presumably could have predicted the 1989 fall of the Berlin Wall!

The Newtonian scheme fostered and encouraged a deep belief in reductionism, which has largely persisted into the twentieth century.[3] Reductionism goes hand in hand with materialism—the belief that inert matter is the source of all phenomena—to provide the prevailing "metaphysics" of classical science. Reductionism, together with materialism, excludes consciousness from any fundamental role in the universe. The objective world exists, is independent of us, and is oblivious of our existence. But we and our consciousness are not independent of it. On the contrary, we are a direct consequence of the inert particles. Newtonian physics obliterated the Cartesian separation of mind and matter. Mind is a product of matter. How does the philosophy of mind reduced to matter fare in twentieth-century physics?

CONSCIOUSNESS IN MODERN PHYSICS

On the surface, relativity and quantum theory seem to return consciousness—in the guise of the observer—to a

3. For a characteristic and popular profession of the modern faith of reductionism, see J. S. Trefil, *The Moment of Creation* (New York: Charles Scribner's Sons, 1983), especially 219-21.

central role in physics. In relativity, the measurement of space, time, and matter is different for different observers. The train and platform observers, as we saw earlier, differ in determining the simultaneity of two events. In quantum theory, Heisenberg's uncertainty principle tells us that an observer has a random and irreducible effect on any observation. Whether an electron acts like a wave or a particle depends on how you observe it.

All of this has rekindled the hopes of some who search for deeper unity and equality between mind and body and who seek more meaning and purpose in the cosmic plan. Advocates of free will, who felt defeated by Newtonian determinism, now feel vindicated by quantum randomness at the heart of matter. Others find a mental structure in the quantum and relativistic worldviews that is reminiscent of the cosmic consciousness of Eastern philosophy and religion. We must examine modern physics more carefully, however, to see whether it is really any friendlier to consciousness than classical physics was.

The role of consciousness in relativity is actually rather superficial. It is not really the observer but her reference frame that determines the outcome of space-time measurements. To speak of an observer in relativity is more or less a linguistic convenience. It does not even imply a live being. For example, an automated observation performed by a computerized telescope on board a high-speed rocket in outer space constitutes a perfectly good observation, which would obey the laws of relativity whether the rocket is manned or not. The theory of relativity does not actually imply that the consciousness of an observer has any effect on measurements of the physical world.

In quantum theory, the situation is more subtle. As in relativity, a quantum observation does not actually require a conscious or live observer. In fact, an observation, according to some physicists, requires nothing more than the recording of information irreversibly in time. Some irreversible process—a chemical change, such as exposing a photographic film to light—is sufficient for an observation whether it's ever seen by a live observer or not.[4]

In contrast to relativity, the choice of the kind of observation has a critical effect on the phenomenon. Choices, however, need not be made by a living being. They can be selected randomly by an automated device from some predetermined alternatives. Some of the EPR test experiments work in this way, with the choice of observing electron waves or particles made arbitrarily.[5]

Furthermore, the act of observation itself, even if performed by a conscious observer, involves no higher function of consciousness such as thinking, imagination, or will. It is pure dumb and blind raw sentience that affects phenomena—a passive act of perception by an unconscious (or, at least, an unself-conscious) mind. An observing worm would presumably have the same effect on an electron as we do. Thus, as far as pure observation in quantum theory is concerned, consciousness is either completely unnecessary or, if it is involved at all, it is only at the most rudimentary level of unconscious sensation.

Incidentally, according to Bohr, quantum theory does not really distinguish between observations and phenomena, as we have already seen. Quantum laws tell us only the results of specific experiments in which the phenomena and the observational apparatus are treated as an inseparable whole. Quantum theory makes no direct statement about a natural world that is independent of our observations. The peculiarities of our language and of our classical concepts make it almost impossible to talk of a synthesized observer-phenomenon. If we take all of this at face value, then we cannot properly draw any conclusions from quantum theory about an independent consciousness. Bohr recognized this as a fundamental obstacle in any

4. See, for example, H. Pagels, *The Cosmic Code* (New York: Simon and Schuster, 1982), 152–54.
5. See, for example, A. Shimony, "The Reality of the Quantum World," *Scientific American* (January 1988): 46.

attempt to extend quantum theory to a broader realm of phenomena, which might include consciousness.[6] But we shall continue to follow the more conventional approach of reasoning about an objective world that is separate from its observers.

There is one final argument—probably the strongest—against thinking that quantum theory might assign any essential role to consciousness. In the last analysis, modern physics is just as materialistic and reductionist as Newtonian physics. Whatever role consciousness may play in nature, its ultimate scientific explanation must be given in terms of the laws governing the behavior of matter. The fundamental elements of all natural phenomena are the elementary particles and their interactions. Life and consciousness are simply complex examples of their activity. Consciousness is no independent force or principle in nature; it is an "epiphenomenon," an unusual but inevitable consequence of the laws of inanimate matter. It may be a difficult and lengthy road to travel, but the explanation of consciousness (as of life itself) will be reduced, sooner or later, to the laws of physics.

Taken as a whole, these arguments and considerations make clear that consciousness is not a welcome guest in modern physics. But is this the end of the matter? Can physics and chemistry ever give a complete and sufficient explanation for life and consciousness? And can science continue indefinitely to deny the fundamental nature of the human experience of mind and consciousness? The scientific rift between spirit and matter is a deep wound in the human heart. It cannot be left unattended.

SUBJECTIVE ELEMENTS IN PHYSICS

Whatever the "official" position of physics toward the human mind may be, there is a great deal of indirect

6. See W. Heisenberg, *Physics and Beyond* (New York: Harper and Row, 1971), 114–15. See also A. Pais, *Niels Bohr's Times, in Physics, Philosophy, and Polity* (Oxford: Clarendon Press, 1991), 438–44.

evidence to suggest more than merely a facilitative role for human consciousness and creativity in science. Many great physical scientists have given accounts of the vital part played by imagination—and even dreams—in their scientific creations—Einstein, Friedrich Kekule, Henri Poincaré, Erwin Schrödinger, Paul Dirac, and Freeman Dyson, to name only a few.

Einstein underwent a profound change of heart in his attitude toward science as the result of the inspiration and insight that led to his general theory of relativity. His youthful advocacy of the empiricist position (the laws of science can logically be deduced simply from the observation of natural phenomena) eventually gave way to his more mature faith in human intuition as the source of scientific theories. As he states it, "Physical concepts are free creations of the human mind, and are not, however it may seem, uniquely determined by the external world."

Friedrich Kekule was an architect turned chemist. His discovery of the ringlike form of the benzene molecule was inspired by a dream he had one night in 1865. In his vision, Kekule saw a snake biting its own tail while whirling in a circle. The benzene carbon ring was the key to the modern structural theory of organic chemistry.

Poincaré traces the choice of "facts" in science to aesthetic criteria of convenience, simplicity, and elegance. Dirac is famous for placing beauty above factuality in evaluating theories of physics. Clearly, there is more to science than just "getting the facts."

There are also philosophers and historians who suggest a more central role for mind and subjectivity in science. Michael Polanyi, a chemist and philosopher of science, argues that there are always tacit elements or unconscious dimensions in scientific knowledge that can never be isolated from the minds in which they originate.[7] Something as simple as using a meterstick or a voltmeter to

---

7. M. Polanyi, *The Tacit Dimension* (Garden City, N.Y.: Anchor, 1967).

make a measurement requires all kinds of skills, familiarities, past experiences—tacit knowledge—that can never be fully communicated or explained and must be passed from one generation of scientists to another through body language, intuition, and imitation. But the tacit dimension permeates all aspects of science—its definitions, concepts, theories, methods, and assumptions.

The physicist and historian Gerald Holton explores the often subjective and irrational motives behind many fundamental ideas in science.[8] Kepler, for instance, resisted assigning elliptical orbits to the planets for many years. In vain he struggled to preserve the pivotal role that the Neoplatonic thinkers assigned to the sun. He could not bear to displace the sun from its position at the center of all the circular planetary orbits. Kepler ultimately capitulated when he realized that the focus of an elliptical orbit plays the same role as the center point of a circle. How many pet beliefs, aesthetic choices, and irrational preferences remain incorporated even today in scientific theories?

There is even a maverick effort to combine mind and matter into a single "hologram" theory. David Bohm, a physicist, and Karl Pribram, a psychologist, have attempted to capitalize on the hologram structure of the mind and the physical world.[9] Both human memory and the quantum description of the universe share a property in common. Every part contains information about the whole, as in a hologram. Memory is not localized but spread out or diffused in the brain. The wave function of an electron gives a very small but finite probability that the electron can be found anywhere in the universe. Thus, all wave functions overlap, and all things refer and relate to each other. In a hologram, unlike in an ordinary photograph, any section can be cut out and used to regenerate the whole image. Each portion of the hologram contains information about the whole. This led Bohm and Pribram to develop a

8. G. Holton, *Thematic Origins of Scientific Thought* (Cambridge: Harvard University Press, 1973).

9. D. Bohm, *Wholeness and the Implicate Order* (London: Routledge & Kegan Paul, 1981).

theory that proposes that mind and matter have the same structure and in reality are different aspects of the same thing.

Finally, we find the physicist Roger Penrose questioning whether a computer will ever be able to do everything that the mind can do. Penrose thinks that Einstein wasn't simply being stubborn when he insisted that quantum theory was incomplete. He concludes that deeper laws than those of quantum theory are needed for the functioning of the mind.[10]

THE EVOLUTION OF CONSCIOUSNESS

The conflict between physics and consciousness is deeper still. Not only may physics be incapable ever of reducing consciousness to the actions of electrons but physics itself, with all its flaws and limitations, may represent a particular stage in the evolution of consciousness. This is the contention of such critics of science as Morris Berman and Theodore Roszak, and especially Owen Barfield.[11]

According to Barfield, the human experience of the world has evolved from an ancient organic participation with nature to our modern objectified separation from it. A study of the language and culture of ancient and primal civilizations suggests an unfamiliar intermingling and fusion of what we think of as the psychic and physical worlds. In Platonic thought, we have already referred to an apparent equivalence between outer and inner space—an equation, as it were, of mental and celestial "motions" that makes little sense to us. Thought, perception, and sensation were forms of motion—*kinesis* in Greek. It was as if the *kinetics* of the mind and of the stars were one and the same.[12]

This Platonic sense of participation gradually

---

10. R. Penrose, *The Emperor's New Mind* (New York: Oxford University Press, 1989).
11. See O. Barfield, *Saving the Appearances* (London: Faber & Faber, 1957).
12. Ibid., 101–3.

dwindled in Western history. Aristotle already singled out for criticism the Platonic "confusion" of space and mind. In the Middle Ages, this spatial metaphysics lingered, if in a weakened form, in medieval astrology, with its characteristic blending of astronomy and psychology that is so incongruous to the modern mind. Today the mind is withdrawn not only from space but from nature itself, as psychology in its most extreme form—behaviorism—denies that mind exists anywhere at all.

All of this, Barfield tells us, reflects an evolution of consciousness rather than merely an evolving conception of the world. Greek and medieval people, as their language, art, and culture suggest, experienced space, time, matter, the heavens, the minerals, and the cosmos differently from the way we do. We describe ourselves and other bodies as objects in an otherwise empty space, like performers on a blank stage. But medieval people thought of themselves more like embryos enmeshed in a cosmic womb, and Plato equates the revolutions of mind in the heavens with the revolutions of the intellect.

It is our consciousness that has changed, as reflected in our language and habitual patterns of thought. To argue conversely that it is our conceptions, rather than our perceptions, that have changed is to reveal the Western scientific belief in an objective independent world.

Neither Barfield's conjecture nor the assumption of science can ever be proven. But if we pursue Barfield's argument, we reach a new interpretation of the "facts." If consciousness has indeed evolved and if experience cannot be divorced from consciousness, then our conceptions and theories amount to mental constructs or representations. The Greeks were well aware of this. Their efforts to "save the appearances" were precisely that. They wanted to reconcile the outward appearances or representations of things with their true inner nature. The world of appearances was inseparable from mind and consciousness. The geometry that could reveal the perfect sphere within the irregular rock was not a theory of reality but a heuristic means to rationalize the mere appearances or representations of things.

With Copernicus, the effort to save the appearances becomes something radically different. The Copernican universe is no longer a model to reconcile the irregular motion of the planets with the perfect harmony of circular celestial motion. It is an actual description of physical reality—of how the planets move in real space.[13] The Copernican revolution not only ousts the earth from a central role in the cosmos, it repudiates our deep connection between mind and matter. It rejects participation.

When we treat our representations as reality, when we forget their original purpose of merely saving the appearances, we commit a profound kind of idolatry. We forget that we created our physical representations and metaphors, and we imbue them with a power and independence that ultimately comes to intimidate and control us. We replace the metaphorical with the literal. We ignore and deny the central role of consciousness in human experience. We transform ourselves from conscious participants in the cosmic pageant to inert puppets, forced to perform a meaningless dance on the barren and empty stage of space-time.[14]

---

13. E. Grant, *Physical Science in the Middle Ages* (Cambridge: Cambridge University Press, 1989), 87.
14. For an alternative, metaphorical approach to physics, see R. S. Jones, *Physics as Metaphor* (Minneapolis: University of Minnesota Press, 1990).

CHAPTER 14

Quantum Theory
Writ Large

DESPITE THE CONFUSION OVER THE meaning and interpretation of quantum theory, its practical consequences and successes have been astonishing. Nowhere is this more apparent than in the study of condensed matter—the science of the solid and liquid states. Nothing is more basic to an understanding of these states of matter than the quantum explanation of how atoms bond together to form molecules. Bonding is responsible not only for such common compounds as graphite and nitrogen gas but also for the symmetric crystalline structure of many metals and gems. By applying quantum theory to these crystals, it has been possible to explain why silver, for example, is a good conductor of electricity and heat but is opaque to light, and why diamond is a poor electrical and thermal conductor but is transparent to light. Of even greater practical significance has been quantum theory's successful explanation of semiconductors, which are the basic materials for transistors and chips, used universally today in electronics. Other triumphs of the quantum theory of large-scale matter have been the invention of the laser and the explanation of superconductivity—a special state of matter in which it conducts electricity almost perfectly.

MOLECULAR BONDS AND FORCES

We have already seen how the new quantum ideas were successfully applied to electrons, photons, and the structure of the atom. Quantum theory continues to provide remarkable insights as we move up the scale of matter to study how atoms combine into molecules. There are two basic types of bonds by which atoms form molecules—ionic and covalent bonds. An atom is normally electrically neutral. It has the same number of positively charged protons in its nucleus as it has negatively charged electrons in captive orbits. (Electrons and protons carry exactly opposite amounts of electrical charge.) An atom can become negatively charged or *ionized* by gaining electrons, which creates an excess of negative charge. Alternatively, if an atom loses electrons, it becomes deficient in negative charge, making it positive. Such charged atoms are called ions. Positively charged ions attract and bond to negatively charged ions to form molecules. This is an ionic bond. For example, when a negative ion of chlorine attracts and attaches itself to a positive ion of sodium, a molecule of sodium chloride—common table salt—is formed.[1]

The details of the process are something like this: When the sodium and chlorine atoms approach each other, an electron is transferred between them. The sodium has one lone electron in its outer state, or shell, as it's often called. It transfers this "loose" electron to the chlorine atom, which is missing just one electron in its outer shell. Sodium is trying to get rid of one electron, and chlorine wants to pick one up. So they strike a bargain. They transfer an electron, become oppositely charged ions, attract each other, and bond into a molecule. (Less figura-

---

1. This is often written as an equation for a chemical reaction: $Cl^- + Na^+ = NaCl$.

tively, it turns out that the bound sodium chloride molecule has a lower energy than the separate sodium and chlorine atoms. A lower energy system is more favorable and stable, and so the molecule forms.)

Ionic bonding, based on the attraction of positive and negative charges, was quite understandable before the advent of quantum theory. But the covalent bond, in which neutral atoms combine into a molecule, was a complete mystery. Two hydrogen atoms, for instance, will form a molecule of hydrogen gas, and several carbon atoms will bond together into molecules of graphite, coal, or diamond. Why should two neutral hydrogen atoms bond into a molecule? Or more precisely, why is a molecule of hydrogen lower in energy, and thus more stable, than two separate hydrogen atoms? This question can be answered only by applying quantum theory to the bonding process.

First, we must recall that a hydrogen atom consists of a single proton—the nucleus—with one electron in a captive state. When two hydrogen atoms approach each other, their combined energy becomes lower if they share their electrons. There are several reasons for this. When the electrons are shared by the two atoms, each electron has more room in which to move around—the space of two atoms rather than one. As we know, the location of an electron is related to its wave function. The wave has a greater amplitude where the electron is most likely to be found. So in the shared state, in which the electron occupies a larger region, the wave function for an electron has a larger amplitude over a greater region of space—it is more spread out. When a wave function is spread out in this way and has a greater extension or stretch, it is said to have a larger wavelength. The larger the wavelength of a wave, the lower is its frequency. (Wavelength and frequency have a reciprocal relationship for all waves.) In quantum theory, a smaller frequency means a lower energy. So, in the shared state of the hydrogen molecule, each electron has a larger wavelength and thus a lower energy than it has in its individual atomic state.

A second reason for the lower energy and greater stability of the hydrogen molecule is that in the shared state the electron waves overlap in the region between the two protons. The two electrons share the ground state of the molecule.[2] This provides a concentration of negative charge between the two positive protons that attracts them and bonds them together. Such a bound state of opposite charges also has a lower energy, just as in the case of the sodium chloride molecule, which has a lower energy than the individual sodium and chlorine atoms. If the two protons in the hydrogen molecule get too close, however, they will repel each other. So there is an optimum distance at which the two protons and the two electrons can live happily ever after. The hydrogen molecule represents the state of lowest energy and greatest stability for two hydrogen atoms.

We see now why the story of covalent bonding could not be written without quantum theory. The wave function of an electron, the relationship of energy to wavelength, and the "overlapping" of electrons are all pivotal to an understanding of the hydrogen molecule. The covalent bond, in contrast with the ionic bond, is purely a quantum effect. As it turns out, this is more the rule than the exception with condensed matter. Many properties of molecules and of matter on the large scale can be understood only through quantum theory. One of the best examples of this is the conduction of electricity and heat.

THE STRUCTURE OF CRYSTALS

It is well known that metals are good conductors of electrical current and heat. The conducting wire in most circuits is made of copper. When electrical contacts are critical in a calculator, say, they are plated with gold—partly because gold doesn't corrode but also because it is such an excellent

---

2. This is allowed by the Pauli exclusion principle, provided that the two electrons have opposite spins so that they are not in identical quantum states.

conductor of electricity. The best cookware is made of copper or aluminum, which can conduct heat more rapidly to the food.

What is less well known about metals is that their atomic structure is crystalline. They have basically the same orderly atomic pattern as precious stones and gems. But diamond and quartz are exceedingly poor conductors of electricity, so much so that they are considered insulators, i.e., materials that block or impede the passage of electrical charge. Why is it that metals are good conductors, while gems are good insulators? And while we're thinking about the contrast between these two crystalline forms, why is it that gems are transparent to light and metals are opaque? Could conductivity and transparency be complementary properties?

A crystal is a giant molecule. Sodium and chlorine atoms do not usually form isolated molecules of salt. Instead, they link together into large arrays of atoms, which are salt crystals on the large scale. The sharing of electrons between atoms occurs either through transfers in ionic bonds or through spreading and overlapping in covalent bonds. Electron sharing can occur among several atoms at the same time, which causes them to link together into large three-dimensional chains or networks, which are called crystal lattices. Some lattices have a cubical structure in which the atoms occupy the "corners" of an endless array of linked cubes, like a giant Tinkertoy. The Tinkertoy "sticks" or edges of the cubes represent the shared electron bonds between the atoms. Other lattices have hexagonal or rhomboidal symmetry, or they may combine several different forms. All crystals exhibit symmetry at the atomic level, which often shows up on the large scale as a symmetric geometrical shape. Because metals are usually melted and cast into specialized shapes, we rarely see the raw crystalline form they exhibit when they are first mined. But both metals and gems are crystals with ordered arrays of atoms.

What determines the conductivity and transparency of

a crystal lattice is the nature of the electron bonds between the atoms. The electrons that are shared among the bound atoms are in lowered energy states, just as in the case of ionic and covalent molecular bonds. This gives the crystal, like the simple molecule, its stability. In the atom, as we have seen, electron states are quantized. An electron cannot occupy any arbitrary position or have any value of energy. The location and energy of an electron is determined by its particular quantum state. The energies take on only allowed discrete values.

In a crystal, the electron states are also quantized, but with a difference. Some of the electrons are shared among all the atoms of the array. Their quantum states are "collective" rather than individual. The shared electrons are not bound to one individual atom but move freely throughout the crystal, held in captive states collectively by all the atoms of the lattice. The electrons in the deeper atomic states—also called inner shells—are strongly attached to their own atoms. These electrons occupy individual states. The electrons in the outer atomic shells are those that are shared in the bonds of the lattice. These "valence" electrons occupy shared or collective states.[3] Such states are not quantized within an individual atom but within the crystal as a whole. Because the collective quantum states are not confined to an individual atom but rather to the dimensions of the whole macroscopic crystal, they have very unusual properties.

The collective states still have quantized energies—their energy values are discrete and separated, not continuous. But the separations, or gaps, between the quantized energies of the collective states are a great deal smaller than the gaps in the individual states, which are confined to the microscopic region of one atom. In other words, the energy levels are far apart in individual atomic states but very close together in the collective electron states of the crystal.

3. Valence electrons were discussed at the end of Chapter 8. They are the outer-shell electrons involved in chemical reactions.

The reason for this is related to the fundamental quantum character of microscopic phenomena, which gradually disappears as we move toward the larger scales of matter. The uncertainty principle states that the quantum nature of an object is inversely related to its size. A tiny electron exhibits a pronounced quantum behavior—particle-wave duality. A planet, however, displays no observable quantum effects—it acts like a well-trained classical particle. The state of an electron in a microscopic atom is dramatically quantized. The discrete energies are few and far between. But the collective electron state for a large-scale crystal is rather weakly quantized: its energies are closely spaced and nearly continuous.

In an atom, each energy state, or shell, is divided into subshells. According to the Pauli exclusion principle, all the electrons in one atomic subshell have the same energy but different spins or rates of orbital rotation. In other words, the electrons in a subshell cannot all be doing the same thing, as it were; they cannot all be in the same quantum state.

What we find in a crystal is that the subshells of the atom still exist. These subshells are now collective states, which occupy the whole macroscopic crystal and involve the interplay of countless atoms. Thus the energy levels in the crystal are not like the separated lines in the atom but instead are separated "bands." Each atomic subshell becomes a compound state, or band, in the crystal, consisting of millions and millions of very closely spaced energy levels. For each subshell in the atom, there is in the crystal a multiple state, or band, with countless "slots" for electrons of very slightly different quantized energies. How the slots in the bands are populated in a given crystal determines its conductivity.

To Conduct or Not to Conduct

The electrons in a collective state of a copper crystal are not bound by the confines of a single copper atom but by

COMPARISON OF BAND LEVELS IN CRYSTALS

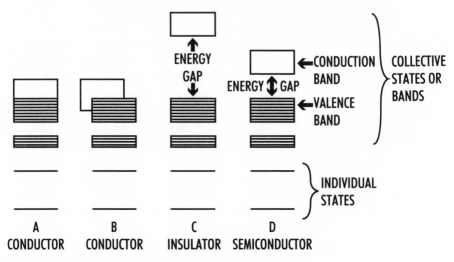

A. CONDUCTOR WITH COMMON VALENCE AND CONDUCTION BAND
B. CONDUCTOR WITH OVERLAPPING VALENCE AND CONDUCTION BANDS
C. INSULATOR WITH LARGE ENERGY GAP BETWEEN VALENCE AND CONDUCTION BANDS
D. SEMICONDUCTOR WITH SMALL ENERGY GAP BETWEEN VALENCE AND CONDUCTION BANDS

the macroscopic dimensions of the whole crystal. The energies of the collective electrons are still quantized and discrete, but they are spaced so closely together as to be effectively continuous. These closely spaced energy levels form an energy band. The various bands, in turn, are separated from each other like the shells and subshells in an atom. Thus the electron states in a crystal form a series of separated bands rather than separated lines. Each band contains countless electrons with a range of energy values that are all but continuous.

The populating of the bands is what determines the properties of the crystal. The higher energy bands of a crystal may or may not be filled with their quota of electrons, just like the shells in an atom. The highest occupied band in a crystal is called the valence band, because it contains the outermost or valence electrons from each atom. If the valence band is not fully occupied with its allowed electron quota, i.e., if it has room for more electrons, then the crystal will be an electrical conductor. If the valence band is full, the crystal will be an insulator.

The highest band that is not fully occupied is called the conduction band. In copper and other metals, the valence and conduction bands are one and the same. In other words, the highest energy band is not fully occupied. (It is both the valence and the conduction band.) The electrons in this band have "room" in which to move around—there are energy values available to them. So any of these electrons in a copper wire can respond with an energy change when a voltage is applied to the wire. This means that the valence electrons in copper can move in response to an applied voltage from a battery or generator. Moving electrons constitute an electrical current. Thus copper is an electrical conductor.

When the valence band is full, however, and well separated from the empty conduction band above it, the material cannot conduct and is an insulator. This is the case for diamond, whose valence band has its full quota and is separated from the empty conduction band by a large energy gap. Each electron in diamond's valence band is prohibited from moving to an already occupied energy level in the valence band because two electrons cannot be in identical quantum states, according to the Pauli principle. Because the band is full, there is no "room" for the electrons to move to other energy values. Furthermore, the electrons cannot "reach" the empty conduction band above the valence band because the much larger amount of energy needed to jump the gap is generally not available. The electrons are stuck. They are unable to respond when a typical voltage is applied.[4] Thus, diamond is an insulator and cannot conduct an electrical current.

4. Many thousands of volts must be applied to break down the resistance of diamond and other insulating materials to force them to conduct electricity.

The same valence electrons, which conduct electricity, also conduct heat. It is the drifting motion of the electrons through the lattice that constitutes an electrical current, while it is the vibration and collisions of the electrons that carry heat through a conductor. Because there are unoccupied energy states in the valence-conduction band, the electrons can conduct electricity and heat. But while copper conducts heat and electricity, it does not "conduct" light: it is opaque. And diamond, which is transparent to light, is a very bad conductor. Why?

The conduction of electricity and heat depends on the ability of electrons to absorb energy from a battery or a hot flame. The conduction electrons can absorb energy because they can move to higher unoccupied energy levels in their band. Thus, the electrons in copper can absorb energy from a variety of sources, including light. Light is electromagnetic energy in transit. It is a traveling wave of electromagnetic energy. A conduction electron in copper can absorb a photon of light and "jump" to a higher available energy level in its band. Since the copper electrons absorb photons, the light wave quickly loses its energy and disappears. The wave does not pass through the copper, and copper is opaque. In diamond, by contrast, the electrons cannot absorb any photons because there are no available energy levels for the electrons to jump to. The light is not absorbed; it passes through the crystal, and diamond is transparent. The "free" electrons in a partly filled band make copper a conductor and opaque. The "fixed" electrons in a full band make diamond an insulator and transparent. A crystal can either conduct or be transparent, not both. Conductivity and transparency do indeed seem to be complementary.

SEMICONDUCTION

There is, however, an in-between case. Certain materials like silicon and germanium are neither full-fledged conductors nor card-carrying insulators. They are semiconductors. The valence band in silicon is full, but the energy gap between the valence and conduction bands is much smaller than in diamond. An electron in silicon can be more easily "coaxed" to conduct because it has only a small gap to jump across. The electrical conductivity of semiconducting materials also improves at increasing temperatures. A little extra available heat energy makes it easier for electrons to jump the gap—so hot silicon is a better electrical conductor than cold silicon. These materials are called intrinsic semiconductors. Their semiconductivity is a natural inherent characteristic.

The big news is that we can create artificial semiconductors. These are called extrinsic or impurity semiconductors because they are produced by impregnating a pure semiconducting crystal with a foreign substance. The process is called *doping*. Typically, germanium is doped with either arsenic or gallium. Now, an atom of arsenic has one more electron than an atom of germanium. When an arsenic atom displaces a germanium atom in a crystal, there is one more electron than is needed in the atomic bonding. This extra electron has new empty quantum states available to it, and it becomes a carrier of electricity. Germanium doped with arsenic has a greater ability to conduct electricity than pure germanium. The amount of conductivity can be controlled by varying the concentration of the doping material. In such an impurity semiconductor, we have a negative carrier of electricity—the extra arsenic electron. Accordingly, germanium doped with arsenic is called an n-type, or *donor*, semiconductor. N refers to the negative carrier and donor means that the impurity atom donates or supplies its extra electron for conduction.

There is also a p-type, or *acceptor*, semiconductor. A gallium atom has one less electron than does a germanium atom. When gallium is doped with germanium, after "the smoke clears" the atomic bonds come up one electron short. This missing electron amounts to a "hole" in the energy band. In quantum theory, this hole is treated like an anti-particle or a positive electron. The hole becomes a positive carrier of electricity. Thus the *p* in p-type. Such materials are also called acceptors because the impurity atom accepts electrons into the hole. When an electron fills a hole, it leaves another hole behind. As successive electrons move to fill the hole, the hole, in effect, moves in the opposite direction. And so the hole progresses through the crystal like a positive charge conducting current.

Semiconduction, by the way, is the basis of the transistor, which was invented in 1948 by John Bardeen, Walter Brattain, and William Shockley, who received the Nobel prize in physics in 1956 for their invention. Because transistors can operate efficiently with very little current and power and because they can be manufactured in small sizes, they rapidly replaced the more cumbersome and expensive vacuum tube in most electrical circuits. The transistor later metamorphosed into integrated circuits and microchips, which have revolutionized the electronics industry.

SUPERCONDUCTORS

Another remarkable quantum property of matter is superconductivity. Among naturally occurring materials, silver, copper, and gold have the greatest ability to conduct electricity. Certain other metals, like lead, tin, mercury, and indium, however, become far better conductors even than silver at very low temperatures. In fact, they become perfect conductors to all intents and purposes. An electrical current, once it is initiated in a ring of superconducting indium, will last for a hundred thousand years, without any boost. Indium, when superconducting, has an effective electrical resistance of zero. The problem with this from the practical point of view is that the temperature at which these materials become superconductors is very near absolute zero—just a few degrees on the Celsius scale (or on the

Fahrenheit scale, for that matter). So it would be quite impractical and terribly expensive to use them in developing superconducting circuits for general use.[5]

The quantum explanation of all of this was given in 1957 in a new theory of superconductivity, called BCS after its discoverers, John Bardeen, Leon Cooper, and Robert Schrieffer.[6] What the BCS theory shows is that at very low temperatures the electrons in certain metals behave in a collective way. This is a pure quantum effect with no classical analogue. In a normal metal, the motion of the free electrons is repeatedly interrupted as the electrons "bump" into the atoms of the crystal. In other words, the crystal lattice itself impedes the motion of the electrons, and this is the source of electrical resistance in a conductor. In a superconductor the electrons act collectively rather than individually. The lattice cannot interfere with any one electron, unless it interferes with all of them at the same time. The chances of such simultaneous interference are negligibly small. Thus, the BCS theory predicts that the superconducting electrons will drift through the lattice without any interference to speak of. This amounts to a current with no resistance—a perfect conduction of electricity, or superconductivity.

In 1986, some new materials were discovered that become superconducting at much higher temperatures than those of lead or mercury—some as high as the temperature

---

5. Superconductivity is used today, however, in medical technology—in magnetic resonance imaging. In addition, building and maintaining superconducting magnets for use in particle accelerators has become a minor industry. The most notable example is the superconducting tevatron at Fermilab near Chicago, which uses a thousand superconducting magnets. The new proposed superconducting supercollider accelerator will require roughly ten times as many superconducting magnets.

6. Bardeen, Cooper, and Schrieffer received the Nobel prize in physics in 1972 for their work in superconductivity. Incidentally, this is the same Bardeen who also won the Nobel prize in 1956 for inventing the transistor. John Bardeen is the only physicist who has ever been awarded the Nobel prize twice.

of liquid nitrogen.[7] This may not sound like a big improvement, but the practical problems of producing and maintaining liquid nitrogen are trivial in comparison with those of maintaining liquid helium, which is needed to cool the older superconductors. The new superconductors are not even metals but ceramic materials—various oxides of copper in combination with such rarer elements as lanthanum, barium, and yttrium.

The search for "hot" superconductors continues. Some researchers conjecture that there may even exist room-temperature superconductors. The possibility of any practical use for the new superconductors is still quite remote. Ceramic materials cannot easily be machined or molded into convenient shapes or extruded into wire. Hot superconductor technology is not just around the corner. Of even greater interest to physicists, however, is the explanation of how the new superconductors work. The BCS theory does not explain them. Quantum theorists have their work cut out for them.

The Laser

The laser is yet another beautiful example of how the microphysics of individual atomic states can be amplified or blown up to produce a collective effect on the large scale of everyday experience. This can occur in certain gases, liquids, and crystals.

The discrete energy levels in an atom represent the stationary states in which an electron can exist without radiating or absorbing energy. When an electron makes a transition or jump from one level to a lower level, a photon is emitted or released. The photon has an energy equal to the difference of energy between the two electron states. The photon's energy, in turn, determines the frequency or color of the photon.

---

7. See, for example, A. M. Wolsky, R. F. Giese, and E. J. Daniels, "The New Superconductors: Prospects for Applications," *Scientific American* (February 1989): 61.

There are three ways in which photons can be emitted or absorbed in an atom. In *spontaneous emission*, an electron drops to a lower energy level and emits a photon. This occurs spontaneously in an atom, according to the probabilistic laws of quantum theory. The probability is different for different states, but in most excited states (states above the ground state), an electron will spontaneously drop to a lower energy state in just billionths of a second.

In *stimulated absorption*, an electron absorbs an incoming photon and jumps from a lower to a higher energy level. The photon has an energy that corresponds exactly to the difference in energy between the levels. By supplying just the correct amount of energy, the photon induces or stimulates the electron to jump to the higher state.

Finally there is *stimulated emission*, in which an incoming photon of just the right energy stimulates an electron to drop down to a *lower* energy state. This may sound strange—why would an electron drop to a lower state when energy is supplied? In the first place, quantum processes are generally reversible and can proceed with equal probability in different directions (as long as certain laws, such as the conservation of energy, are not violated). Secondly, in stimulated emission, the energy of the photon corresponds to the difference between the energy of the electron's state and that of a *lower*, rather than a higher, state. It's as if the vibrating photon induces a sympathetic vibration in the electron that can make it go up or down. Of course if the electron is stimulated to drop down to a lower energy level, then energy is released rather than absorbed. A new photon is produced, with exactly the same energy and frequency as the first one, and so two identical photons are emitted. These two photons are also exactly in phase with each other, i.e., they vibrate together in unison. The two photons are also monochromatic (they have the same color) as well as coherent (in phase).

Stimulated emission is the basis of the laser, which is actually an acronym for light amplification by stimulated emission of radiation. The laser was developed in the 1950s by Charles Townes, Nikolay Basov, and Aleksandr Prokhorov, who received the Nobel prize in physics for their work in 1964. Laser action occurs as a kind of snowball effect when there are many electrons in the same excited atomic state in certain gases, liquids, and crystals. One photon of the right frequency stimulates one electron to drop down, thus producing two identical photons, which in turn induce two more electrons to drop down, producing four identical photons, and so on. This chain reaction results in the production of countless identical photons, which combine into a light beam.

To make a laser, several practical problems must be solved. First, how can we get many electrons to stay in the same excited state? Second, how do we trigger the laser process? Third, how do we form and control the light beam?

The first problem requires achieving a *population inversion*—inverting, as it were, the electron population from the ground state to an excited state. The problem is largely solved by nature itself, which fortunately provides some atomic states with a very low probability of *decaying*, or making a transition to a lower state. These states are called *metastable*. Typically, they last for thousandths, rather than billionths, of seconds. Once electrons are in a metastable state, they will remain there long enough for the laser chain reaction to occur.

A common type of laser uses a mixture of helium and neon gas. The gas atoms are excited by a high-voltage electrical discharge, which causes them to move around rapidly and bounce into each other. The voltage is chosen so that when two atoms collide there is a very good chance that a helium atom will boost a neon atom into a certain metastable state. Because the neon electrons are "stuck" in this state for some thousandths of a second, the laser action has time to occur. The action is triggered when any electron in a metastable neon atom spontaneously drops to the ground state and thus emits a photon of just the right energy to start the ball rolling.

The helium-neon gas mixture is contained in a long and thin cylindrical glass tube whose end faces are silvered to reflect the laser light. The photons that travel along the length of the cylinder keep reflecting back and forth and contributing to the chain reaction. Photons traveling in other directions simply pass through the transparent cylindrical walls of the tube and are lost. Thus, an intense parallel beam of light is produced traveling along the length of the cylinder. One end of the tube is only partially silvered, and it allows some of the beam to emerge each time it reflects. The laser light emerges in the form of an intense narrow beam of light with very little spread. It is monochromatic (red, in the case of a helium-neon laser) and coherent, which gives it a uniquely characteristic grainy and patterned appearance.

Because laser beams spread out so little and are so intense, they can travel great distances without much attenuation. They have been successfully reflected from the moon! The precise directionality and linearity of laser beams make them extremely valuable in applications that require precision measuring, monitoring, and surveying. The intensity of the beam determines how much energy it carries and how much heat it can produce when it is absorbed in matter. A laser beam can pinpoint a heat spot well enough to weld, drill, or cut metal. Extremely high-energy lasers are being used in experimental research, which attempts to use nuclear fusion as a practical energy source.[8] These fusion lasers can focus enough energy on hydrogen to raise its temperature to the millions of degrees required for hydrogen fusion to occur.

Laser beams of lower energy can be focused so narrowly and accurately that they are used in delicate surgical procedures, such as reconnecting a detached retina. A whole new medical industry is based on laser light and optical technology. Laser beams can also be modulated to carry information in a very compact form, like an electri-

---

8. See Chapter 16.

cal current. But laser light traveling through thin transparent filaments can carry information far more efficiently and compactly than current can, as it travels through metal wires. Optical circuitry and information systems are already in use in computers and the recording industry. Optical output channels have become commonplace on compact-disc players. Laser technology is destined to generate whole new industries, which unfortunately already have a destructive by-product in the form of an x-ray laser weapon for use in some future star wars. Quantum theory has given us the amazing potential of the laser. What ultimate use shall we make of it?

CHAPTER 15

The Atomic Nucleus

SOON AFTER LAUNCHING THE SUCCESSFUL quantum
theory of the atom, physicists began to explore the atom's
tiny massive core—the nucleus. The positively charged
nucleus attracts and holds in captive states the negative
electrons in the atom. But what holds the nucleus itself
together? The nucleus consists of positive protons and
uncharged neutrons. The protons repel each other with an
enormous repulsive force (which the uncharged neutrons
do not feel). What holds the nucleus together and over-
comes the mutual repulsion of the protons is a new force—
the *strong force*—that operates only within the nucleus. It
is this powerful nuclear force that is responsible for the
enormous energy of an atomic bomb. The study of the
properties of atomic nuclei and nuclear forces has profoundly influenced the twentieth century. Radioactivity,
isotopes, nuclear reactions, fission, fusion, atomic energy,
nuclear weapons, and nuclear medicine are all by-products
of nuclear physics.

The Atom's Heart of Darkness

In 1911, Ernest Rutherford proposed the planetary model of the atom with its electrons orbiting a tiny but massive core—the nucleus—concentrated at the center of the atom. For the next two decades, most research in physics focused on the outer electronic structure of the atom. This work established quantum physics as the new theory of the microworld and laid the foundation for the remarkable applications of quantum theory to large-scale matter. But the tiny nucleus at the heart of the atom remained a mystery.

James Chadwick's discovery of the neutron in 1932 was the key that opened the door to the modern age of nuclear physics. Despite the minute size of the nucleus (one-ten-thousandth the size of the whole atom), it is a complex structure that can be broken down into simpler components—neutrons and protons. The proton was already known as the simple nucleus of the hydrogen atom, but the neutron was a new actor on the nuclear stage. The neutron is only slightly heavier than the proton and is electrically neutral, as its name suggests. The neutron is found in the nuclei of all atoms except that of the simplest form of hydrogen. Because there are as many protons in the nucleus as there are electrons in the atom, the proton number determines the character and chemical properties of the atom. The neutrons, on the other hand, seem to affect the stability of the nucleus—so we know that the neutrons and protons play different roles in the nucleus. The big question was what held the neutrons and protons together in the first place.

The positively charged protons in the nucleus are so close together that they exert an enormous force of repulsion on each other. What keeps them from flying apart? Only the existence of a much stronger nuclear force—separate and distinct from the electrical force of repulsion—could explain the integrity and stability of the nucleus. The energy associated with this strongest of all of nature's forces was ultimately to fuel the reactor and the bomb. (We'll explore atomic energy in Chapter 16.)

As it turns out, there are two disparate nuclear forces. The so-called strong nuclear force is indeed responsible for holding the nucleus together (and also for spawning the bewildering horde of elementary particles that lurk deep within the nucleus). A separate *weak* nuclear force presides over the instability, rather than the cohesion, of the nucleus. The weak force controls a particular kind of spontaneous breakdown or radioactivity of the nucleus, which is called beta decay. This is a decay or decomposition of the nucleus, which is accompanied by the emission of electrons, or beta rays, as they're called in the context of radioactivity.

NUCLEONS AND THE STRONG FORCE

The inhabitants of the nucleus—the neutron and proton—are collectively referred to as *nucleons*. Nuclei are made of nucleons. The term *nucleon* is used because the strong nuclear force does not distinguish between neutrons and protons but treats them essentially as the same particle. The electrical charge of the proton relates only to the electrical force. All nucleons carry a nuclear charge, as it were, which is independent of any electrical charge. This nuclear charge is the source of the strong attraction between nucleons, just as an electrical charge is the source of the electromagnetic repulsion between protons. The strong nuclear force is a separate and independent force that acts between nucleons.

The number of protons in the nucleus is called the atomic number, Z, and is the same as the number of electrons in the atom. The neutron number is N. The number of nucleons, A, is simply the sum of N and Z.[1] There are 92 basic nuclei, one for each of the 92 atoms, and they are commonly listed in the same order as that of the elements of the periodic table. The "lightweight" nuclei near the beginning of the list generally have equal numbers of protons and neutrons (N = Z). For example, the nucleus of

1. A is usually called the atomic mass number.

helium-4 has 2 protons and 2 neutrons, boron-10 has 5 of each, and silicon-28 has 14 of each.[2] As we move toward the "heavyweight" nuclei, we find there are always more neutrons than protons. Gold-197 has 79 protons and 118 neutrons, and uranium-238 has 92 protons and 146 neutrons. The preponderance of neutrons over protons for heavyweight nuclei is related to nuclear stability and the unique character of the strong nuclear force, as we shall see.

The strong nuclear force, although significantly stronger than the electromagnetic force, has a much shorter reach or range. The range of the electromagnetic force, like that of the gravitational force, is infinite. Two charges or two masses anywhere in the universe exert forces on each other. These forces are very weak when the charges or masses are great distances apart, but they are never zero. Because there is some small force, no matter how vast the distance, the range of the force is said to be infinite. By contrast, the strong force has an extremely short range— about the diameter of a nucleon. The strong attractive nuclear force, which holds the nucleons together, acts only between adjacent nucleons. This is because nonadjacent nucleons are farther apart than the diameter of a nucleon; they are beyond the reach of the strong force. If there are three nucleons in a row, the strong force reaches from nucleon one to nucleon two, and also from two to three, but not from one to three. The strong force acts only between nearest neighbors in the nucleus.

Incidentally, the range of the weak force is much shorter than the range of the strong force by about a factor of 1,000! A comparison of the four fundamental forces of nature is summarized in the chart on the next page.

Because the strong force acts only between adjacent nucleons, all protons in the nucleus repel each other via the electromagnetic force, while only adjacent protons (and neutrons) attract each other via the strong force. The neu-

---

2. The number attached to the element name is the nucleon number, A. Helium-4 has four nucleons—two protons and two neutrons.

Force/Field	Particles Affected	Relative Strength	Range
Gravity	All matter	1	Infinite
Electromagnetism	Charged particles	10^{36}	Infinite
Weak force	Neutrons, protons, electrons	10^{29}	10^{-18} meters
Strong force	Neutrons, protons	2×10^{37}	10^{-15} meters

trons, which experience nuclear attraction but not electrical repulsion, moderate the effect of the protons. Extra neutrons in the nucleus, which provide attraction without repulsion, help "dilute" the disruptive repulsion of the protons. The more protons there are, the more neutrons there must be to keep the nucleus stable. This explains the preponderance of neutrons over protons in large stable nuclei. Beyond a certain size, the cumulative effect of all the repelling protons begins to exceed the strong attraction between only neighboring nucleons, and the nucleus becomes unstable. All nuclei beyond bismuth, the eighty-third nucleus, are unstable. They spontaneously disintegrate or decay into smaller nuclei. Beyond the ninety-second nucleus, uranium, nuclei are so unstable that they do not exist naturally. Plutonium, the ninety-fourth element and the most poisonous substance known, is artificially created by earthlings, primarily for the production of atomic weapons. It is found nowhere else in the universe. Nature knew better than to make plutonium.

NUCLEAR SIZE AND MASS

Much of the research in nuclear physics has centered on the effort to understand the strong nuclear force, which is difficult to observe directly because of its very short range. Important evidence has come indirectly from studying the size and structure of nuclei, which are determined by the strong force.

A nucleon is about one-trillionth of a millimeter in diameter. (This is also about the range of the nuclear force.) By comparison, an atom is one hundred thousand times larger—about a ten-millionth of a millimeter. The mass of a nucleon is about one-trillionth of one-trillionth of a gram.[3] The size and mass of a nucleus is determined by the number of nucleons it contains. A nucleus of oxygen is roughly sixteen times as massive as one of hydrogen, because oxygen nuclei contain sixteen nucleons (eight neutrons and eight protons) and hydrogen nuclei only one (a proton).

The diameter of an oxygen nucleus is only two and one-half times that of hydrogen (and not sixteen times). This is because the nucleons aggregate in the nucleus like marbles in a spherical bowl. A sphere of sixteen identical marbles has only two and one-half times the diameter of one marble. The largest nucleus—uranium, with 238 nucleons—has a diameter about six times larger than hydrogen's. The diameters of all the nuclei fall evenly into place as we move from lightweight hydrogen to heavyweight uranium. This provides an important clue about the nuclear force.

In contrast to nuclei, the sizes of atoms do not increase uniformly as we move through the periodic table. Atomic electron states do not fill up like marbles in a bowl. Electrons are spread far apart in the mostly empty space around the nucleus, according to the rules for quantized electron states. Successive electron states are not simply related to each other in size, so that adjacent atoms in the periodic table may be significantly larger or smaller than each other.

Nucleons, within the close quarters of the nucleus, act like hard balls in contact. Nucleons cannot penetrate each other, and the distances between nucleons are the same in all nuclei. Thus, as the number of nucleons increases, the size of the nucleus simply reflects the "close packing" of the

---

3. It is impossible to conceive of these sizes. Describing them quantitatively is extremely useful in science, but we must be careful to take these numbers with a grain of salt, lest we fall prey to the folly of the quantitative that we discussed earlier.

nucleons. The density or compactness of all nuclei is the same, unlike that of atoms. What this means is that nucleons repel each other at very close ranges. The strong force causes nucleons to attract each other when they are roughly a nucleon diameter apart, or a little less. But when they get too close—within a small fraction of a nucleon diameter—they repel each other. The nuclear force is complex. It has an ultrashort-range repulsive component, as well as a short-range attractive one. This explains both the uniform density and the relative stability of nuclear matter.

ISOTOPES

There are ninety-two varieties of nuclei—one for each atom of the ninety-two chemical elements. But, in fact, there are many more than ninety-two nuclei. The nucleus of chlorine has seventeen protons, but there are two kinds of chlorine nuclei—one with eighteen neutrons and one with twenty neutrons. Both nuclei, chlorine-35 and chlorine-37, are stable and found in nature.[4] The chemical properties of chlorine are determined by the number of protons (or electrons), which is its atomic number, $Z = 17$. Since chlorine-35 and chlorine-37 each have seventeen protons, they are chemically indistinguishable. They are called *isotopes* of chlorine, i.e., nuclei with the same number of protons and different numbers of neutrons (same Z, different Ns). From the atomic or chemical point of view, they are identical, but they have different nuclear structures that reflect the peculiar vagaries of the strong force and nuclear stability.

As it turns out, there are quantized nucleon states in the nucleus, just as there are quantized electron states in the atom. These states are also populated according to the Pauli exclusion principle, so there are characteristic patterns and repeating periods among the nuclei, just as there

4. Chlorine-35 has thirty-five nucleons ($A = 35$) and chlorine-37 has thirty-seven nucleons ($A = 37$).

are among the atomic elements. Certain combinations of neutrons and protons are more favorable and stable than others. Chlorine-36 with nineteen neutrons is highly unstable and does not naturally occur, while both chlorine-35 and chlorine-37 are found in common table salt. The last two are the stable isotopes of chlorine. Tungsten, with seventy-four protons, has five stable isotopes. Thus, there are several hundred different nuclei, because many elements have several isotopes. Deciphering the pattern of stable and unstable isotopes and their natural abundances has also contributed significantly to the puzzle of the nuclear force.

Radioactivity

In 1896, Henri Becquerel, a French physicist, discovered that certain materials emit a form of radiation that is not affected by chemical change or the absorption of light. In other words, this radiation has nothing to do with the electronic states of the atom and atomic spectra. The penetrating radiation originates in the spontaneous breakdown or decay of a nucleus, which is said to be *radioactive*. The precarious balance of nuclear attraction and electric repulsion makes certain nuclei unstable (usually, but not exclusively, the heavyweight nuclei). These nuclei get rid of their excess destabilizing energy by transforming themselves into more stable, less energetic nuclei. In the process, radiation is emitted in one of three forms, called *alpha, beta,* or *gamma rays*. While still unknown, these forms of radiation were named for the first three letters of the Greek alphabet, and they were later identified more specifically. Alpha rays are high-speed helium nuclei (two neutrons and two protons). Beta rays, as we already mentioned, are electrons. Gamma rays are high-energy photons, even higher in frequency and energy than x-rays.

Among these naturally occurring radioactive processes, we discover examples of the amazing "transmutation" of elements, long sought after by medieval alchemists. For example, a nucleus of lead can spontaneously decay into a nucleus of bismuth by emitting an electron or beta ray. The emitted electron comes from a spontaneous transformation inside the nucleus: a neutron changes or decays into a proton plus an energetic electron. This change results in a net increase of one in the number of protons in the nucleus (the atomic number increases by one), which causes a transformation of lead into bismuth—the next higher element in the periodic table.

Radium "transmutes" itself into radon by emitting alpha rays. This radium-alpha decay, by the way, is the culprit responsible for excess radon gas that is found in some buildings. The small amount of radium in building materials decays continuously into radon, which is a radioactive gas that may mix with the air in your home. If there is not good air circulation, as is often the case in basements, radon can accumulate in potentially dangerous amounts. Radon, unfortunately, is both colorless and odorless and can be detected and monitored only with special sensitive devices.

Gamma decay does not result in any transmutation, but it gives clear evidence of the quantized nucleon states inside the nucleus. An excited barium nucleus, for example, will spontaneously emit a high-energy gamma ray and "drop" to the ground level. This is exactly analogous to the case of a hydrogen atom that emits a photon when an electron drops from an excited state down to the ground state. Now we are talking of nucleon states inside the nucleus rather than electron states in the atom. The energies of the nucleon states are much greater than electron energies in the atom (typically a million times greater). So gamma rays emitted from nuclei are much more powerful and penetrating than the photons of atomic spectra. The process is the same, however, and nuclear spectra reflect the quantized states of neutrons and protons in the nucleus.

DECAY RATES

The moment at which any particular radioactive nucleus will decay cannot be predicted with any certainty. Like all processes in the microworld, radioactivity is described only by the statistical laws of quantum theory. Regular decay patterns can be observed only for large numbers of radioactive nuclei by using the law of averages, just as life-expectancy tables can be applied only to whole populations and never to individuals. What we do observe is that certain radioactive materials (samples containing countless atoms) decay more rapidly than others. Characteristic decay rates vary greatly—from trillionths of a second to billions of years.

The time it takes for half of a radioactive sample to decay is called its *half-life*. Radium has a half-life of 1,620 years. If I start with a gram of radium, then after 1,620 years, I (or more properly, my descendants) will have one-half gram of radium left. The other half gram will have decayed into radon. After another 1,620 years (3,240 years altogether), there will be only half of a half, or one-quarter gram of radium left. In each half-life of 1,620 years, half of the radium sample will disappear.

It will take forever, or at least a very long time, for all of the radium to decay. But the 1,620-year half-life gives me some indication of the decay rate. A bottle of muons with a half-life of two-millionths of a second will not last very long. Strontium-90, a particularly dangerous and virulent form of radioactive fallout, has a half-life of twenty-nine years. Unfortunately, once it is produced, it hangs around for centuries. Uranium-238 plods along more or less eternally with a half-life of 4.5 billion years. Decay rate and half-life are also measures of the instability of the nucleus. They add yet other pieces to the puzzle of the nuclear force.

NUCLEAR REACTIONS

Although the nucleus is held together by the strong force, we see that other forces are involved in the stability and decay of nuclei. Beta decay, for example, is mediated by the weak force, and alpha and gamma decay by the electromagnetic force. In an effort to understand the interplay of these three forces—strong, weak, and electromagnetic—as well as the general structure of the nucleus, physicists artificially induce other kinds of nuclear transformations. These are called nuclear reactions, and generally they involve collisions or interactions among nuclei and elementary particles. An example would be a collision between an alpha particle and a nitrogen-14 nucleus. In this reaction, a sample of nitrogen-14 is exposed to a radioactive source of alpha rays. The alpha bombardment transforms the nitrogen into oxygen-17 through the release of a proton. We shall study this reaction in a little more detail to learn how it reveals information about the nucleus.

The alpha particle, you will recall, is a helium nucleus. It contains two protons and two neutrons—four nucleons in all. Nitrogen-14 contains fourteen nucleons—seven neutrons and seven protons. When the alpha particle strikes the nitrogen nucleus, there is a momentary combination of the eighteen nucleons of the two initial particles. This is a "compound" nucleus of nine neutrons and nine protons, which is highly unstable. The compound nucleus immediately decays into a stable oxygen nucleus with seventeen nucleons—eight protons and nine neutrons. The leftover proton is emitted and carries off the excess energy. We can summarize the nucleon interchange as follows: $4 + 14 \rightarrow 18 \rightarrow 17 + 1$.

The arrows represent the steps in the process as they occur successively in time, but the arrows also act effectively as equal signs. In fact, this example illustrates a fundamental characteristic of nuclear reactions, namely, that the number of nucleons is preserved. We have the same number of nucleons—eighteen—before, during, and after the reaction. You may have noticed that the number of neutrons and protons was separately conserved in this reaction, which automatically guarantees the conservation of nucleon number. But this isn't always the case. The number of nucleons is *always* preserved, but not necessarily the number of neutrons and protons. In the lead-to-bismuth beta decay mentioned above, all the nucleons are preserved, but one neutron becomes a proton and emits an electron. Bismuth actually has one more proton than lead, because one of the lead neutrons turns into a proton.

This transformation of a neutron into a proton and an electron is typical of many beta-decay reactions. But notice this: Although we "lose" a neutron and "gain" a proton, there is no net loss or gain of nucleons. Nor do we have any net gain or loss of electrical charge. The neutron is neutral, and the combined charge of the proton and electron is zero—also neutral. Both the number of nucleons and the amount of electrical charge are conserved in this and in all nuclear reactions.

The strong force preserves nucleon number. It does not distinguish neutrons from protons. Neutrons and protons may come and go, but the total number of nucleons is rigidly maintained. In these reactions electrical charge is also strictly preserved. Positive and negative charge may appear, but only in equal pairs whose net charge is neutral. Neutrons may turn into protons, lead into bismuth, and nitrogen into oxygen, provided that the number of nucleons and the amount of charge is strictly preserved—or conserved, as we say in physics. This salient feature of nuclear reactions—the conservation of certain quantities during a transformation or change—has become a central and seminal idea not only in nuclear physics but also in the modern understanding of elementary particles and their interactions. Nuclear physics ultimately led the way to our deepest understanding of matter through the introduction of such conservation laws and symmetry principles as we see operating in these simple nuclear reactions.

NUCLEAR MEDICINE

Radioactivity and nuclear reactions have had important applications outside of physics. An example is the field of nuclear medicine, in which tracer amounts of radioactive materials are used to diagnose and treat certain ailments.[5]

---

5. For more information on nuclear medicine, see M. K. Loken, "Nuclear Medicine: Past, Present and Future," *Journal of the Association of Physicians of India*, vol. 26, No. 2 (February 1978): 61.

Starting with its discovery in 1896, radioactivity has been used continuously for medical applications. In the early years, only radium and its related by-products were used for therapeutic purposes. After World War II, new radioactive materials, which were produced by atomic reactors, became available. This marked the start of the modern era of nuclear medicine, which is characterized by the use of new "designer" radioactive materials that can pinpoint malfunctioning body organs for diagnosis or treatment.

Tracer amounts of radioactive substances (i.e., minute quantities that can be detected by their radiation but which have a negligible effect on human tissue) are administered to a patient intravenously. These *radiopharmaceuticals* are organic materials, which accumulate in particular body organs that use them to function. Iodine, for example, collects in the thyroid gland, diphosphanate in the bones, albumin in the blood, and sulfur colloid in the liver. The molecules of these radiopharmaceuticals are modified, or *labeled*, by attaching to them a radioactive atom, usually technetium-99m, which does not affect the chemistry of the molecule. (Ninety-nine is the nucleon number of technetium, and the *m* stands for metastable.) Technetium-99m has a half-life of about six hours and emits gamma rays. Thus, within a day or two, all traces of this material are gone from the body. But during the first few hours after the material is administered, it accumulates in an organ and emits gamma rays that can be continuously detected and used to monitor the functioning of that organ. A special *scintillation* camera, connected to a computerized display, is used to detect the gamma rays from the organ. These cameras use a scintillation crystal of sodium iodide, which emits light, or scintillates, when it absorbs gamma rays.

A patient with a heart problem, for example, will be given a radiopharmaceutical that collects in her blood. The patient then lies on a table with the scintillation camera suspended above her chest. On a nearby display screen, the doctors will see a live televised image of the blood flowing through her heart. Because the radiopharmaceutical is in the body of the blood itself, the image is not of the heart muscle but of the actual flow of blood through the chambers and arteries of the patient's heart. The pattern and rates of blood flow can be observed and measured over a period of time and used to make a precise diagnosis of any problem.

This technique has an advantage over alternative, more invasive procedures. In an angiogram, for example, a catheter is inserted through a blood vessel directly into the heart. Then a *contrast* material is administered into the heart so that special x-rays may be taken to locate arterial blocks. The catheterization, the contrast material, and the x-ray are all far more invasive than the administration and monitoring of a radiopharmaceutical. Furthermore, the detection and monitoring of the technetium gamma rays allows for a quantitative measurement of the blood flow, which is not possible in an angiogram.

These techniques of nuclear medicine are used far more for diagnosis than for therapy. Apart from the treatment of specialized diseases of the thyroid and blood, there are few "atomic bullets" that have been used successfully. Radiopharmaceuticals and instrumentation have been designed for monitoring and diagnosing the function of many body organs, including the brain, liver, spleen, kidneys, and lungs as well as blood and bones.

In nuclear medicine, the potentially dangerous rays of radioactive substances have been harnessed and tamed for the benefit of humankind. Unfortunately, the applications of nuclear fission and fusion, which we'll discuss in Chapter 16, have been far more problematic.

Atomic Energy
and the Bomb

WE DO NOT YET HAVE A COMPLETE quantum theory of
the nuclear force, but we understand enough to have made
possible the practical development of atomic energy. The
key has been the study of the pattern of stability and instability of nuclei, with an eye to Einstein's conversion of mass
into energy. Fission and fusion are the nuclear reactions by
which heavy and lightweight nuclei are transformed into
more stable middleweights. These two processes can unleash the tremendous power of the nuclear force and make
it available for atomic bombs and nuclear reactors.

The energy locked up inside the nucleus has become
the bane of the twentieth century. The promised use of
atomic energy as a cheap source of domestic and commercial power has never fully been realized, as the by-products
and failures of atomic reactors continue to plague us. The
only really "successful" application of atomic energy has
been the development of modern nuclear weapons, whose
ability to inflict unimaginable destruction threatens the
very survival of humanity and all life on earth.

A NEW SOURCE OF ENERGY

During the 1930s, it became clear to physicists and chemists that there is a huge untapped store of energy locked within the atomic nucleus. Because the force that holds the nucleus together is so strong, the energy associated with this force is enormous. When the vast energy of a nucleus is released, it consumes only a tiny amount of nuclear matter. A very small quantity of matter can be transformed into an immense amount of energy, as is reflected in Einstein's famous formula, $E = mc^2$. The complete conversion of one gram of matter into energy, as we saw in Chapter 1, would produce enough heat to boil the water in eighty-five Olympic swimming pools.

For most chemical reactions, like the burning of coal and oil, the decrease in mass, which is the quantitative measure of matter, is much too small to be observed. The enormous energy of a nuclear reaction, however, can be detected as a loss of mass. The energy associated with the strong force that binds the nucleus together actually represents a few tenths of one percent of the mass of the nucleus. The practical problem was how to unleash, and yet control and moderate, this spectacular conversion of matter into energy. Two natural processes—nuclear fission and fusion—were discovered that would make nuclear energy a twentieth-century reality.

NUCLEAR STABILITY

In Chapter 15 we noted a kind of competition in the nucleus between nuclear attraction and electrical repulsion. Because of the short range of the strong nuclear force, only adjacent nucleons attract, but all protons electrically repel each other. For large nuclei, the cumulative electrical force can significantly weaken the cohesion of the nucleus and make it unstable. This is forestalled in heavyweight nuclei, to some extent, by the presence of more neutrons than protons. The extra neutrons add nuclear attraction without electrical repulsion.

But there is a limit to the increase of neutrons over protons. All nucleons in the nucleus are in quantized states, populated according to the Pauli exclusion principle. Neutrons and protons each occupy their own separate states. As extra neutrons are added to the nucleus, they must occupy states of ever higher energy, and this excess neutron energy also tends to make the nucleus unstable. Thus, there is a limit on the neutron number. As we move to larger nuclei, the problem eventually gets out of hand. Among the heavyweights, the larger nuclei are less stable, and most nuclei beyond uranium are so unstable that they exist only for fleeting instants when artificially produced by humans.

You might think, therefore, that the smaller a nucleus is, the more stable it is. Nothing is quite so simple in nuclear physics. Lightweight nuclei are also less stable than middleweights. Recall our picture of the nucleus as a bunch of nucleons in a ball, like marbles in a round bowl. The nucleons near the "surface" of the ball have a greater chance of escaping from the nucleus, just as the molecules in a drop of water escape by evaporation more easily from the surface of the drop. The nucleons at the surface of the nucleus (like the molecules at the surface of a water drop) are less strongly bound to the other nucleons. This is because surface nucleons are not surrounded on all sides by adjacent attracting nucleons. If a surface nucleon chances to have a little extra energy (and everything in quantum theory is a matter of chance), it can escape. Thus nuclei with a large percentage of their nucleons near the surface are likely to lose nucleons, and this makes them unstable.

It is a mathematical fact that a small ball of marbles has a larger percentage of marbles at the surface than does a large ball. If you form a "ball" with two, three, or four marbles, then clearly they're all at the "surface." There are no "interior" marbles, and you have 100 percent surface marbles. As you add marbles, you begin to get interior marbles, which are completely surrounded by other marbles. Now only a certain *fraction* of the marbles is at the surface, and therefore the percentage of surface marbles is less than 100. As the number of marbles increases, a larger and larger fraction of the marbles is located in the interior of the ball, and the percentage of surface marbles decreases. With a smaller percentage of marbles at the surface, a large ball is less likely to lose its marbles than a small one. (It's not only people who lose their marbles!)

Nucleons and water molecules aren't marbles, of course, but they all share the same geometrical proportion or ratio of surface area to interior volume. A large sphere has proportionately less surface than a small one, as contradictory as it may sound. (The emphasis here is on the proportion or ratio.) In a small nucleus, as compared with a large one, there is a greater chance of losing nucleons through surface "evaporation," which makes lightweight nuclei relatively less stable.

In summary, electrical repulsion tends to make heavyweight nuclei unstable, and surface "evaporation" tends to make lightweight nuclei unstable. Thus the most stable nuclei are the middleweights, which are neither too large nor too small. Since unstable systems tend to strive after stability, there exist natural "pathways" (nuclear reactions), by which heavyweight and lightweight nuclei are transformed into middleweights. Heavyweights becoming middleweights is called fission. Lightweights becoming middleweights is called fusion.

THE MASS-DEFECT CURVE

In almost any book on nuclear physics, you will find a diagram of the so-called mass-defect curve.[1] What this curve illustrates is the relative stability of the nuclei. It is a graphic picture of how the excess energy of nuclei varies with their size or nucleon number, A.

Stable nuclei have lower energies on the average than

---

1. It's also called the curve of binding energy and the packing-fraction curve. See, for example, I. Asimov, *Understanding Physics: The Electron, Proton and Neutron* (New York: New American Library, 1969), 186. See also J. McPhee, *The Curve of Binding Energy* (New York: Ballantine, 1976).

MASS-DEFECT CURVE

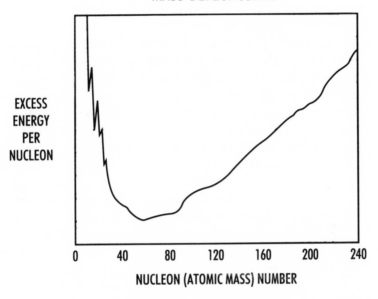

EXCESS ENERGY PER NUCLEON

NUCLEON (ATOMIC MASS) NUMBER

0   40   80   120   160   200   240

unstable ones of comparable size. Excess energy is an indication of instability. Lightweight nuclei, like deuterium (an isotope of hydrogen with one neutron and one proton) or helium with low nucleon numbers, have excess energy because of the evaporation effect. Deuterium and helium aren't really unstable; they are simply relatively less stable than middleweights, like carbon and oxygen. Heavyweight nuclei, like thorium and uranium with large nucleon numbers, have excess energy because of the electrical repulsion effect. They are relatively less stable than middleweights, like gold and krypton.

Large nuclei have more total energy simply because they have more nucleons. To get a measure of relative stability, we use the energy per nucleon and not the total energy of the nucleus. Carbon-12 has roughly three times as much energy as helium-4 because it has three times as many nucleons. But the energy *per* nucleon of carbon and the energy *per* nucleon of helium are almost the same. They would be exactly the same if carbon and helium were equally stable, but because carbon is more stable than helium, its nucleons have less energy on the average than those of helium. So the energy per nucleon is slightly less for carbon than for helium.

The mass-defect curve shows how the energy per nucleon varies with nucleon number. It's called a mass-defect curve because lower energies (greater stabilities) show up as a decrease or defect in mass, as we discussed earlier. Lightweight nuclei, with low values of A, and heavyweight nuclei, with high values of A, all have large average energies per nucleon. Middleweight nuclei have smaller average energies per nucleon. So when you plot a graph of energy per nucleon against nucleon number, the curve is high at the ends and slopes down in the middle, like a very loose clothesline. The middleweight nuclei with A-values around fifty are the most stable. They have the least energy per nucleon and lie along the lowest part of the curve. Larger and smaller nuclei, when given a chance, tend to "fall" toward the low middle region of the curve, just like beads at the ends of the clothesline would tend to slide toward the middle.

In other words, in a nuclear reaction a heavyweight nucleus will typically be transformed into middleweight nuclei. To do this, and also to conserve the number of nucleons as is required in all reactions, heavyweights tend to split apart into more stable middleweights. This is the process of *fission*, which means splitting. By the same token, lightweight nuclei in a nuclear reaction will also be transformed into more stable middleweight nuclei. Doing this, while conserving nucleons, requires lightweights to combine or fuse into middleweights. This is the process of *fusion*. So the relationship of nuclear stability to size, as illustrated by the mass-defect curve, explains the natural processes of nuclear fission and fusion, which have given us atomic energy.

NUCLEAR FISSION

The mass-defect curve tells us that the nuclei at either end of the periodic table are relatively the least stable and have the highest energy per nucleon. Any nuclear reaction that transforms heavyweight or lightweight nuclei into middle-weights lowers the average energy of the nucleons and liberates the excess energy. The energy left over in the reaction is released in the form of explosive power and radiation. Splitting large nuclei apart into smaller nuclei—nuclear fission—is one of several reactions that achieve nuclear stability. More commonly, large nuclei like ura-nium are gradually transformed, through a series of beta and alpha decay reactions, into a lighter, more stable nu-cleus like lead. Fission reactions are less likely and gener-ally need to be induced by some external agent.

The uranium-235 isotope is sufficiently unstable that it doesn't take much to induce its fission or splitting. If a neutron strikes and enters a nucleus of uranium-235, the newly formed compound nucleus of 236 nucleons is highly unstable. It rapidly breaks apart, usually into two smaller nuclei, or *fission fragments*, and several leftover neutrons. In the process a great deal of energy is released. Many different middleweight isotopes may show up among the fission fragments of uranium. In a typical fission reaction, a bombarding neutron enters a uranium-235 nucleus and causes it to split into the isotopes barium-144 and krypton-89. In addition to the fission fragments, three neutrons are emitted. The large amount of energy released in the reaction gives the fragments and neutrons high velocities and also produces penetrating gamma rays.

Notice that the nucleon number is preserved in this reaction: 1 neutron + 235 uranium nucleons = 144 barium nucleons + 89 krypton nucleons + 3 neutrons. Barium and krypton have lower neutron-to-proton ratios than ura-nium, and so three neutrons are left over in this reaction. Different fission fragments will occur with the splitting of each uranium nucleus, but invariably a few extra neutrons are emitted. In a sample of uranium-235, these extra neu-trons can strike other nuclei and cause them to split, result-ing in a chain reaction. The fission of one uranium nu-cleus can cause two or three others to occur, and each of these will induce several more, so that the reaction will grow exponentially. This naturally occurring chain reaction in fission is the secret of atomic energy. Once fission is triggered in a large mass of uranium-235, it will rapidly snowball into an uncontrolled nuclear explosion.

This chain reaction depends on the speed of the neutrons (slow ones are better at inducing fission in U-235) and also on the chances that a neutron will randomly strike a uranium nucleus. The fission of uranium-235 naturally produces slow neutrons, so it is much more *fissionable* than the far more abundant uranium-238 isotope.

To increase the likelihood that emitted neutrons will strike uranium nuclei, you must insure that the nuclei are closely spaced and well surrounded by other nuclei. Because of the high density of uranium, this requirement is easily satisfied by using a sufficiently large and compact mass of the material. In such a *critical mass* of uranium, the fission neutrons have a high probability of striking a nucleus, and the chain reaction rapidly proceeds to an explosion. The critical mass for uranium-235 is roughly thirty pounds; for the even more fissionable plutonium-239 it is only about five pounds.

NUCLEAR FUSION

At the other end of the periodic table, lightweight nuclei also have excess energy, because of their higher relative instability. This energy can be released by the fusing of small nuclei into larger ones. An example is the fusion of hydrogen nuclei into helium. The reaction proceeds through several steps in which protons—hydrogen nuclei—fuse into higher isotopes, first of hydrogen and then of helium. The upshot is that four protons fuse into a helium-4 nucleus (two neutrons and two protons) and emit two positrons (positive antielectrons), along with gamma-ray photons and other particles. Here again we see that the four nucleons are preserved, and that the electrical charge is conserved when two protons are transformed into two neutrons and two positively charged positrons.

In contrast with fission, which can be triggered by a stray neutron, a fusion reaction is extremely difficult to induce. For two protons to attract each other and fuse, they must be no farther apart than the range of the strong nuclear force. To get that close, they must overcome the electrical repulsion between them, which becomes quite enormous as the distance between the protons decreases.

The situation is very different in the case of fission. A neutron carries no electrical charge and has no difficulty approaching and penetrating a positively charged uranium nucleus. But the only way for two protons to get very close is by approaching each other at extremely high speeds. If they are moving rapidly enough, then even though their mutual repulsion slows them down, they will still have enough speed to approach closely enough for the strong force to take hold. Thus the fusion of protons requires that they be moving at very high speeds.

In practice, to make protons in a sample move rapidly, you must raise the temperature of the sample. The trouble is that the temperature required to trigger hydrogen fusion is about ten million degrees Celsius (20 million degrees Fahrenheit). While such temperatures are commonplace in the interior of the sun and other stars, they are exceedingly difficult to attain and control on the earth. Such temperatures are indeed achieved on earth in the uncontrolled explosion of a hydrogen bomb, which uses the heat of a fission bomb to "fire" the hydrogen fusion reaction. It is no simple matter, however, to achieve the necessary temperatures under the controlled conditions required in a fusion reactor. It is because nuclear fusion demands such enormous amounts of heat that it is called a *thermonuclear* reaction.

THE ATOMIC PILE OR FISSION REACTOR

The most common fission reactor for producing atomic energy in the United States is of the light-water type. This reactor uses ordinary water (as opposed to heavy water) to moderate or slow down the neutrons produced in uranium fission and to absorb the heat of the reaction.[2] The reactions take place in the shielded isolated core of the reactor. The core typically consists of a stack or pile of tens of thousands of fuel rods, weighing about 100 tons. The rods contain "enriched" uranium, in which the percentage of the uranium-235 isotope has been artificially increased to a few percent. The amount of uranium used in the pile far exceeds the critical mass for fission, and so a set of special control rods, which easily absorb neutrons, is inserted into the pile. The control rods can temper the activity in the core, adjust the rate of fuel consumption, and shut down the reactor in the event of an emergency.

The energy of the neutrons and radiation produced in the reaction shows up primarily as heat. Water under high pressure is used both to cool the reactor core and to transfer the energy to the outside. The heated pressurized water circulates from the core to a heat exchanger, where it transfers much of the heat to a separate system of pipes in which water is vaporized to run a steam turbine. Typically, these reactors generate about 3,000 megawatts of heat, of which 1,000 megawatts is actually converted into electrical power in the turbines. The cost to build such an installation is about one billion dollars.

During the course of the reactor's operating life, there is a gradual accumulation of radioactive fission products in the fuel rods. After a year or so at full power, the accumulated radioactivity in the core amounts to about ten billion curies, or the equivalent of ten thousand tons of radium. Ultimately, this nuclear waste must be stored in a safe isolated location for tens of thousands of years to prevent it from harming the earth and its inhabitants. The long-term threat of reactors is not only from some dramatic meltdown

2. The molecules of heavy water consist of an oxygen atom and two atoms of deuterium, rather than hydrogen. Deuterium, an isotope of hydrogen, contains a neutron in addition to the usual proton in its nucleus. Because deuterium is twice as heavy as hydrogen, the water formed with deuterium atoms is called heavy water.

and explosion but also from the contamination of the environment through the accidental release of vast quantities of radioactive materials.

Estimating the chances for various kinds of reactor accidents is very difficult. We have some information from such catastrophes as Three Mile Island and Chernobyl. But the precise combination of human error and technical failure that characterized those two cases was not anticipated, and the probability for such events is very difficult to estimate. In evaluating the future of the fission reactor, there is more to be concerned with than its potential for poisoning the environment and the problem of the long-term storage of nuclear wastes. The reactor continues to play an essential role in the manufacture and proliferation of nuclear weapons. This ultimately might be the reactor's most menacing consequence.

THE FUSION REACTOR

Although it is far more difficult to induce fusion than fission, the temptations to create a practical fusion reactor are very great for three reasons. First, fusion is more efficient at producing energy. The fusion of a pound of hydrogen releases nearly ten times as much energy as the fission of a pound of uranium. Second, there are fewer radioactive by-products from fusion than from fission. Third and foremost, the supply of hydrogen is almost limitless on the earth, as compared with the relatively small amounts of uranium in the earth's crust.

Unlike the case of the fission reactor, however, a practical reactor for fusion is a long way from being realized. The problem is how to heat, pressurize, and contain matter under conditions approximating those at the center of the sun. It has proven exceedingly difficult, to say the least. Two basic schemes have been under investigation for several decades.

The first employs magnetic fields to guide and confine a hot gas of ionized protons called a *plasma*. The plasma is produced by heating hydrogen gas until the atoms are ionized, or stripped of their electrons. This gas of moving charged particles—the plasma—can be controlled by a magnetic field, which exerts forces on electrical charges in motion.

No solid material container could possibly survive the enormous temperatures required for fusion. But huge magnets of various shapes and designs can be used to confine the rapidly moving protons in a kind of magnetic "bottle." The method sounds fine in principle, but unanticipated and poorly understood instabilities in hot plasmas have continued to plague this approach. The "tokamak" doughnut-shaped Russian design seems to offer the greatest promise and has received the strongest support. A commercially viable plasma reactor of this type, however, is still far in the future.

The second scheme is known as inertial confinement. It is based on the law of inertia—the tendency of stationary matter to remain at rest and resist acceleration. The idea is to zap a tiny pellet of deuterium and tritium (the heavy isotopes of hydrogen that are easiest to fuse) with so much energy from all sides at once that the pellet will be compressed and heated sufficiently to trigger a fusion reaction. The vast power required is focused on the pellet by twenty or thirty awesome laser beams, which momentarily combine to produce a level of power higher than that of the entire U.S. energy grid. Immediately after this incredible bombardment of energy, inertia will hold the matter of the pellet stationary for the few billionths of a second that it takes for fusion to occur.

This technique requires an exceedingly accurate alignment of the laser beams, which are monitored and controlled by a large battery of microprocessors. It is also necessary to manufacture nearly perfect spherical pellets, whose tiny irregularities can be seen only with an electron microscope. Yet, despite the amazing technical feats of the project, it has remained very difficult to demonstrate the scientific (let alone the practical) feasibility of fusion through inertial confinement. The technique tends to consume far more energy than it produces, but research to improve its efficiency is ongoing.

In recent years, some scientists claim to have discovered "cold fusion," i.e., a natural fusion reaction that takes place at room temperature and does not require the enormously high temperatures that characterize the plasma and inertial techniques for fusion. The cold fusion process, if authentic, would offer a very simple inexpensive method to generate almost unlimited quantities of energy by fusing deuterium, which is very plentiful in seawater. The process, however, has not been verified or clearly reproduced by other scientists, and there is considerable doubt and skepticism about the validity of the claim.[3]

THE BOMB

The original pure-fission atomic bomb was essentially a device that could assemble several pieces of pure uranium-235 or plutonium-239 into a critical mass, trigger a fission reaction with neutrons, and keep it all together long enough to allow a chain reaction to escalate to large-scale explosive proportions. The later version—the hydrogen bomb—is based on fusion. It uses a fission explosion, or rather implosion, to heat, contain, and ignite a core of fusion fuel. In contrast to the unattainable controlled fusion reaction, the H-bomb produces an uncontrolled explosion. The modern bomb combines both fusion and fission and incorporates their best and worst features—the greater explosive force of fusion with the virulent radioactive products of fission.

Today's bombs are typically manufactured in the one to ten megaton range, although larger "mammoth" and smaller "strategic" weapons are stockpiled as well. A megaton bomb is one that can produce the explosive force of a million tons of TNT. The original fission bombs, dropped

---

3. See *Physics Today* (June 1989): 17.

on Hiroshima and Nagasaki, were twenty-kiloton bombs, with only one-fiftieth the explosive force of a megaton bomb.

The immediate effects of a nuclear detonation are heat, shock, and radiation. The thermal radiation from the fireball of a one-megaton bomb can cause third degree burns at a distance of seven or eight miles from ground zero. Under certain circumstances, it can also produce a horrendous fire storm, such as the conflagration in Hiroshima after the atomic attack.

The shock wave from the airburst of a megaton explosion generally combines with its own reflection from the ground to form a blast wave of enormous power. Within three miles of ground zero, most commercial buildings and factories would be flattened. Moderate and light damage to buildings and residences would occur out to distances of ten miles. People not close enough to be vaporized by the fireball but within five miles of the blast would be either fatally burned or crushed to death by falling debris and collapsing buildings. The mechanical effects of the blast itself are the most predictable and calculable, and it is primarily these that are used to estimate and evaluate the military effectiveness of a bomb.

The prompt radiation from a nuclear detonation primarily takes the form of neutrons and gamma rays. They are pretty well attenuated within a mile or two of the explosion and do not cause much damage beyond this distance. At closer range, the radiation dose is enormous and highly lethal, but most people in this zone would already have been killed by the blast or the fire.

If a one-megaton bomb were ever detonated a mile above New York City's Empire State Building, essentially all the people within four to five miles of the famous landmark would be killed. If weather conditions were right, a massive fire could break out that would extend the lethal zone to ten miles. This would destroy the entire population of New York City proper.

Delayed and long-term effects of the bomb are much more difficult to estimate. A detonation at or near ground level produces countless radioactive nuclei. These are the result of nuclear reactions caused when the bomb irradiates the matter on the ground. Then, as the ground material initially vaporized by the fireball condenses, it becomes impregnated with these radionuclei. The radionuclei also attach themselves to particles of dust and dirt that are whipped up by the strong winds following the blast. All of these contaminated particles eventually descend to the ground as radioactive fallout. The extent and severity of fallout effects thus depend on the height of the detonation above the ground as well as on the prevailing winds and weather conditions in the days and weeks after the blast. Radiation burns and sickness, delayed cancers, and genetic defects are among the possible consequences of fallout.

The loss of most medical and sanitation facilities and of food and fresh water supplies are impossible to predict accurately but would take a severe toll. Bombing of large areas of the earth's surface could have serious and irreversible effects on the environment and the weather and might sufficiently disturb the ozone layer of the atmosphere to allow the ultraviolet radiation at the earth's surface to reach perilous levels.

Other unanticipated effects of nuclear war undoubtedly exist. A few decades ago, the electromagnetic pulse phenomenon was discovered. A nuclear blast will radiate a pulsed high-voltage wave of electromagnetic energy that can disrupt and permanently damage all transistorized, solid-state, and microelectronic devices. This would incapacitate most cars, radios, television sets, computers, telephones, and power systems.

All of this, though fortunately unrealized, constitutes our legacy from atomic energy. Humankind can never escape from the awareness and vigilance that this inheritance demands.[4]

---

4. For further information on atomic bombs, see K. N. Lewis, "The Prompt and Delayed Effects of Nuclear War," *Scientific American* (July 1979): 53; and J. Schell, *The Fate of the Earth* (New York: Avon, 1982).

CHAPTER 17

Physics and Conscience

THE INVOLVEMENT OF SCIENCE IN today's political and governmental affairs has become a fact of life. The superpower nations support vast industrial technologies based on twentieth-century physics. Scientific advisors sit on the highest councils of government. The principal support for scientific research and development and for reform in science education comes from government funding. The collaboration of government, science, and technology contributes to and aggravates many of our environmental, legal, and cultural crises.

The ethical problems raised by the science-government connection are thorny and complex. Ultimately they all depend on the moral attitudes and decisions of individuals. We will get some sense of this by examining the careers of two great scientists and their relationship to government affairs. Albert Einstein, who left Germany just before the Nazis came to power in 1933, later brought to the attention of President Roosevelt the need for the United States to develop an atomic bomb lest Germany develop one first. He suspended his pacifist views for the only time in

271

his life—to fight Hitler. J. Robert Oppenheimer was the father of the atom bomb and directed its development during World War II. Later he was falsely accused of disloyalty to the United States and was denied further access to classified information.

In the last analysis, our personal attitudes and unconscious beliefs about science amount to moral choices and acts. If we blindly accept the pronouncements of scientists and surrender our right to participate in political decisions involving science, if we worship and idolize science and treat it as a state religion, and if we accept the "scientific method" as the final arbiter of truth and human experience, then we truly abandon our moral and ethical responsibility as human beings.

SCIENTIFIC ACTS OF CONSCIENCE

Most people today feel powerless to influence the decisions and acts of governments, tribunals, corporations, and religious institutions. The proliferation of nuclear weapons, the escalating pollution of the environment, the insane policies of nations, the control of information, the dictates of fanatic nationalistic religions, and the curtailment of human freedom and activity all seem beyond the control of the individual, even in the most democratic of societies and groups. And yet, the decisions and acts of all social bodies are ultimately conceived of and implemented by individuals. Despite the mob psychology of the twentieth century, the moral choices and courageous acts of individuals, such as Mohandas Gandhi and Martin Luther King, Jr., continue to persuade and inspire us, and occasionally even to influence governments.

In the area of physics, no single development has created a greater moral crisis for twentieth-century humanity than the discovery of nuclear energy and the invention of the atomic bomb. In an effort to humanize the moral dilemma of the nuclear age, we shall now look briefly at the case histories of two scientists whose personal moral choices had a profound effect on the development and ultimate uses of atomic energy and weapons. In studying these individual responses to the perplexing dilemma of the atom, we can perhaps come to understand how we as individuals can help influence for the better the atomic age of the future.

ALBERT EINSTEIN[1]

After the verification of general relativity in the eclipse measurement of 1919, Einstein became a world-famous figure. Although puzzled and bemused by his sudden fame, Einstein found a use for his celebrity status. He had lived in Berlin through World War I, which had fully confirmed and justified his early childhood revulsion to arbitrary authority and militarism. Einstein became committed to pacifism and began to use his unique position as both a popular and intellectual leader to espouse his political philosophy. He took advantage of his new role as world traveler and lecturer to crusade on behalf of both world government and Zionism. His first trip to the United States in 1921 was to raise money for Palestine.

During the 1920s in Germany, several bigoted scientists attempted to discredit Einstein's physics as "Jewish" and "communist." These efforts failed for the most part because of Einstein's unassailable position as the leader of the community of physicists and because of his obvious high moral idealism and humanitarianism. But the political climate in Germany and in Europe generally was deteriorating. Einstein withdrew from participating in the League of Nations because of its ineffectiveness, and he personally denounced the gains made by the Nazis in Germany.

During Einstein's visits to England and the United States in 1931 and 1932, the grounds were laid for his later emigration. After personally witnessing Nazi policies in

---

1. An excellent biography of Einstein, which includes discussion of his work, is *'Subtle Is the Lord . . .'* by A. Pais (New York: Oxford University Press, 1982).

Germany, Einstein modified his pacifist views and began to advocate the arming of Europe as an essential step to preserve human freedom.

In December of 1932, Einstein left Germany for a visit to the United States. Soon thereafter, Hitler ascended to power in Germany. Einstein renounced his German citizenship, as he had done once before, and never again returned to his homeland.

As a young high school student, disgusted with the Prussian military style of German schools, Einstein had renounced his citizenship and left Germany to join his family living in Italy. For a while he had been a man without a country, until he settled in Switzerland and became a Swiss citizen. Much later, after he had become a world figure, he was appointed as director of the Kaiser Wilhelm Institute for Physics in Berlin. He had reluctantly accepted German citizenship once again—but only jointly with his Swiss citizenship. A number of German physicists, including Werner Heisenberg, cooperated with the Nazi regime. Einstein never forgave his former colleagues, except for Max von Laue, who became an outspoken critic of Nazi oppression and worked to protect German Jews.

Returning from America to England in 1933, Einstein was warned that it would be too dangerous to reenter Germany. His important papers and a few personal effects were secreted out of the country, but his home and private possessions were all confiscated by the Nazis. Einstein had escaped by the skin of his teeth. He accepted a position as a foundation member of the Institute for Advanced Study in Princeton, New Jersey, where he lived for the remainder of his life. Eventually, he became an American citizen.

In 1939, Enrico Fermi, one of many illustrious European physicists who had emigrated to America, informed Einstein of the recent work on uranium fission in Germany by Hahn, Strassman, and Meitner, and of the possibility of the making of an atomic bomb, based on a chain reaction in uranium-235. After experiments in the United States established that such a bomb was indeed feasible,

Leo Szilard, a Hungarian emigré physicist, and others prevailed upon Einstein to exert his influence by writing a letter to President Roosevelt. In the letter, Einstein advocated developing a U.S. atomic bomb as a precautionary measure in case Germany might develop such a bomb. After receiving a second letter in 1940, Roosevelt acted swiftly to establish the Manhattan Project, which was placed under the leadership of General Leslie R. Groves of the U.S. Army and J. Robert Oppenheimer, a leading American physicist (whose case we shall discuss below).

Einstein took no part in the effort to develop the bomb. (In any case, he was considered a "security risk.") His main concern was that Hitler should not be the sole possessor of this weapon. Einstein was shocked by the Hiroshima bombing in 1945, and he worked strenuously in the final decade of his life to prevent the future development and use of the bomb. He counseled the establishment of a sovereign world government to control the bomb and maintain peace. After World War II, he resumed his pacifist stance and maintained it to the end of his life. He justified his advocacy of military resistance against Hitler as a necessary and rational exception. As he put it himself in a letter to a Japanese correspondent in 1953:

> I am a *dedicated* but not an *absolute* pacifist; this means that I am opposed to the use of force under any circumstances, except when confronted by an enemy who pursues the destruction of life *as an end in itself.*[2]

One week before his death on 11 April 1955, he signed a joint manifesto with the English philosopher and pacifist Bertrand Russell that remains as valid and as wise a statement on atomic energy as on the day it was written:

---

2. B. T. Feld, "Einstein and the Politics of Nuclear Weapons" in G. Holton and Y. Elkana, eds., *Albert Einstein—Historical and Cultural Perspectives* (Princeton: Princeton University Press, 1982), 376.

Here then is the problem which we present to you, stark and dreadful and inescapable: Shall we put an end to the human race; or shall mankind renounce war? People will not face this alternative because it is so difficult to abolish war.

The abolition of war will demand distasteful limitations of national sovereignty. But what perhaps impedes understanding of the situation more than anything else is that the term *mankind* feels vague and abstract. People scarcely realize in imagination that the danger is to themselves and their children and their grandchildren, and not only to a dimly apprehended humanity. They can scarcely bring themselves to grasp that they, individually, and those whom they love are in imminent danger of perishing agonizingly. And so they hope that perhaps war may be allowed to continue provided modern weapons are prohibited.

This hope is illusory. Whatever agreements not to use the H-bomb have been reached in time of peace, they would no longer be considered binding in time of war, and both sides would set to work to manufacture H-bombs as soon as the war broke out, for, if one side manufactured the bombs and the other did not, the side that manufactured them would inevitably be victorious.[3]

If anything, Einstein and Russell understated the case. Atomic stockpiles were developed in the absence of war. Einstein retained his optimism despite the nuclear race he witnessed, but he had his occasional moments of ironic pessimism:

If I were a young man again and had to decide how to make a living, I would not try to become a scientist or a scholar or teacher. I would rather choose to

---

3. Ibid., 386–87.

be a plumber or a peddler, in the hope of finding that modest degree of independence still available under present circumstances.[4]

A fellow physicist, Bernard T. Feld, has well summed up Einstein and his deep moral commitment to humanity:

In reviewing his life, what is overwhelmingly impressive is the degree to which he regarded himself as a servant of his fellow human beings.[5]

J. Robert Oppenheimer[6]

Oppenheimer was born in New York City on 22 April 1904 of wealthy German immigrant parents. At Harvard, he excelled in Latin, Greek, physics, and chemistry, published poetry, and studied Oriental philosophy. He was graduated in 1925, did research at the Cavendish Laboratory in England under Rutherford, and attended Göttingen University in Germany, where he met Bohr, Max Born, Dirac, and other prominent physicists. After receiving his doctorate in 1927, he returned to the United States to take an academic position, first at the University of California at Berkeley and then at the California Institute of Technology, where he did research in subatomic physics and went on to train a whole generation of American physicists who were greatly affected by his qualities of leadership and intellectual independence.

During the mid-1930s, Oppenheimer sided philosophically and morally with the Republicans in the Spanish Civil War. His wife and some of his friends were Communists. He flirted briefly with the Communist movement but ultimately rejected it as parochial and immoral, especially

---

4. Ibid., 385.
5. Ibid., 389.
6. An excellent account of Oppenheimer's role in the development of the atomic bomb can be found in R. Rhodes, *The Making of the Atomic Bomb* (New York: Simon and Schuster, 1986).

after hearing about Stalin's maltreatment of scientists.

In August 1942, General Leslie R. Groves organized the Manhattan Project to construct an atomic bomb and appointed Oppenheimer to direct the entire effort. Oppenheimer established his laboratory at Los Alamos, near Santa Fe in New Mexico—an isolated and favorite area of Oppenheimer's where he had often vacationed in his youth. The project hired a remarkable team—the cream of American, British, and European expatriot scientists—all motivated by the common desire to beat Germany in building the bomb. Apart from its grim goal, the work on the bomb was experienced by the Los Alamos scientists as a dream come true. For three uninterrupted years, the best scientists in the world worked on the most challenging and stimulating science in the vanguard of research and knowledge. Furthermore, the work was carried out with the maximum possible technical and financial support. Oppenheimer frankly described the experience as "sweet."

In the last months of the project, the work became very intense and proceeded at an exhausting pace. V-E Day—the World War II victory in Europe—went by almost unnoticed at Los Alamos. Against some mild opposition from Oppenheimer, a "reconsideration" forum was called to question the continuing of work on the bomb in the light of Germany's defeat. Oppenheimer's view—that, in any event, the world needed to know and appreciate what the bomb was—prevailed, and the work continued. On 16 July 1945, the world's first atomic bomb was exploded at a test site near Alamogordo, New Mexico. Despite anticipating this event for three years, Oppenheimer and the others were thunderstruck by the awesome power and beauty of the explosion.[7]

Einstein and Leo Szilard attempted to intercede with Roosevelt to stop the use of the bomb on Japan, but Roosevelt died on 12 April 1945. Military efforts to mine Japa-

---

7. For a dramatic description of the Alamogordo test, see Rhodes, *The Making of the Atomic Bomb*, 672–76.

nese harbors, which seemed destined to end Japan's ability to resist, were halted for being too "cowardly." Diplomatic efforts to end the war on the part of the Japanese were obstructed by the Russians and by Truman, who had succeeded Roosevelt as president. Szilard's attempts to circulate a petition among the scientists against the use of the bomb were frustrated by Groves, and the petition was treated bureaucratically by Truman. Ultimately, the popular idea prevailed that using the bomb on Japan would end the war sooner and save American lives, although this rationale has become more questionable and controversial in retrospect. A uranium-235 bomb was dropped on Hiroshima on 6 August 1945, and a plutonium-239 bomb was dropped on Nagasaki on 9 August 1945. Japan surrendered a few days later. The momentum of events was so great, according to the American physicist Freeman Dyson, that it is unlikely that anyone could have stopped the use of the bomb.[8]

After the war, Oppenheimer became a top advisor to the U.S. government and worked hard to stop the further production and military use of atomic bombs. From 1947 to 1952, he was chairman of the General Advisory Committee of the Atomic Energy Commission. In 1949 this committee took a position opposing the development of the hydrogen bomb. This did not go over well with the powers that be.

In 1953, at the height of the McCarthy era of government suspicion and loyalty oaths, Oppenheimer was accused of having associated with Communists and of opposing the hydrogen bomb. In a security hearing he was declared not guilty but was denied any further access to classified information. Damaging testimony was given at the hearing by Oppenheimer's former friend and colleague, Edward Teller, the father of the hydrogen bomb. In the estimate of Oppenheimer's brother Frank and also of Isi-

---

8. See R. Jungk, *Brighter than a Thousand Suns* (New York: Harcourt, Brace, Jovanovich, 1958): 208-9.

dore Rabi, a Nobel laureate in physics and a personal friend of Oppenheimer's, this condemnation ultimately destroyed him. He passed the remaining years of his life at Princeton and died of throat cancer in 1967.

But despite his early advocacy and leadership in developing the bomb, Oppenheimer came to see the bomb as a turning point in history—as a new opportunity for world peace. In his farewell speech in 1946 as director of the Los Alamos Laboratory, he said:

> . . . [W]hen people talk of the fact that this is not only a great peril, but a great hope, this is what they should mean . . . there exists a possibility of realizing, of beginning to realize, those changes which are needed if there is to be any peace.
>
> Those are very far-reaching changes. They are changes in the relations between nations, not only in spirit, not only in law, but also in conception and feeling. . . . [A]tomic weapons are a peril which affects everyone in the world, and in that sense a completely common problem, as common a problem as it was for the Allies to defeat the Nazis.

In another talk in 1946, Oppenheimer said:

> It did not take atomic weapons to make war terrible. It did not take atomic weapons to make man want peace, a peace that would last. But the atomic bomb was the turn of the screw. It made the prospect of future war unendurable. It has led us up those last few steps to the mountain pass; and beyond there is a different country.

THE MORAL DIMENSION

Despite their different roles in the development of the atomic bomb, Einstein and Oppenheimer both came to understand the need for a unified moral position in which the needs and desires of humanity took precedence over those of individual nations. Then they went on to work and advocate on behalf of their positions.

Many other physicists also took a stance about the atomic bomb: Leo Szilard tried to prevent the use of the first atomic bomb and later strenuously opposed the development of the hydrogen bomb. In stark and ironic contrast with Szilard was Edward Teller. Both were brilliant Hungarian physicists with very similar backgrounds and educations. Both emigrated to the United States and became involved in the Manhattan Project. But while Szilard became a passionate opponent of all atomic weapons, Teller went on to become the father of the hydrogen bomb, to repulse the views of both Szilard and Oppenheimer, and to caution and warn the U.S. government against them.

Werner Heisenberg was one of the great German physicists who remained in Germany during the war, ostensibly to provide leadership for the younger German scientists and because he believed he could thwart the development of an atomic bomb in Germany. Heisenberg established the Uranium Club in Germany during the war. He claimed that he wanted to convince the authorities that the construction of a bomb was a near practical impossibility and that they should concentrate instead on the development of atomic power.

Unfortunately, in recent years, the role of Heisenberg in Nazi Germany has come under question. Evidence has come to light to suggest that while Heisenberg did not openly cooperate with German authorities, his relationship with them was ambiguous and not as morally pure as he has claimed.[9] Nevertheless, after World War II, Heisenberg actively promoted the peaceful use of atomic power, and in 1957 he led other German scientists in opposing

---

9. See M. Walker, "Heisenberg, Goudsmit and the German Atomic Bomb," *Physics Today* (January 1990): 52. For a good personal account of Heisenberg's life, work, and philosophy, see W. Heisenberg, *Physics and Beyond* (New York: Harper & Row, 1972).

a move to equip the West German Army with nuclear weapons.

Even when the issues seem clear, there will be dramatic differences of opinion on the morality of scientific applications and technology. A serious new ethical problem has arisen in the 1980s and 1990s over the matter of genetic engineering and recombinant DNA and the possible interference with the evolution of species, including humanity. Opinions conflict both inside and outside of the biological community.

But an even deeper dilemma results from the moral neutrality of those who either choose to ignore the consequences of scientific research or refuse to recognize the ethical implications that underlie the very assumptions of science.

SCIENTIFIC IDOLATRY REVISITED

The scientific worldview, which includes our most deeply held beliefs about nature, existence, space, time, and matter, has prevailed in Western civilization and its derivatives for the past several centuries. Western science has been so successful and so dominant in the world that individuals and societies tend to lose sight of its human roots and character and to forget that science is based as much on assumptions, biases, and beliefs as on "facts." Attributing to science an authority and power that ignores its human origin and character is a modern form of idolatry whose moral consequences threaten and endanger humanity and the planet, as we saw in Chapter 7.

Some contemporary physicists, when advocating their latest TOEs—or theories of everything—display a modern hubris by suggesting that these theories are indeed the final authoritative description of the physical world. (We've heard such claims before.[10]) Nevertheless, most scientists

---

10. D. M. Greenberger, letter in *American Journal of Physics* (February 1988): 82.

freely admit that science has limitations when they think seriously about it and discuss it among themselves. In their day-to-day research, in their teaching and lecturing, and especially in their public statements, however, it's quite another matter. Furthermore, statements by science popularizers and journalists often sound as if the theories are actual and unquestioned descriptions of reality.

In a *Newsweek* article on brain chemistry, all the creative output and lofty thoughts of humanity are attributed to "an interaction of chemicals and electricity inside the labyrinthine folds of the brain . . . [which is] something with the consistency of Jell-O and the color of day-old slush."[11]

In a typical popularization of the big bang, the author touts the ultimate triumph of reductionism and castigates some "physicist who broke faith with the reductionist philosophy and began to look outside of the Western tradition for guidance . . . [in] a more holistic, mystical view of nature." He goes on to deny any connection between "nature" and the state of mind of the observer and to proclaim that the modern theories of physics "tell us that the correct description of nature is that description in which the observer is irrelevant. . . . Presumably, it will take a while for this realization to percolate away from a small group of theoretical physicists and become incorporated into our general world view."[12] No false modesty here. Science has already found the "correct" description of nature, and we had all better make the best of it.

The claim that quantum theory does really maintain the separation of mind from nature was in fact argued in Chapter 13. But the question is whether quantum theory should have the final word on the matter. Is the severing of mind from nature in quantum theory—and in all of phys-

11. *Newsweek*, 7 February 1983, 40.
12. J. S. Trefil, *The Moment of Creation* (New York: Scribner, 1983), 219-21.

ics since Newton, Galileo, and Descartes, for that matter—really anything to brag about?

Idolatry is immoral fundamentally because it is a deliberate act of ignorance and denial. If we insist that not only the earth's plants and animals but its very rocks, water, and air are completely separate and independent from us—from our minds and our consciousness—then they are not subject to our consciences either. American Indians treat all aspects of nature as holy, filled with spirit, and connected to the mind and soul of humanity. No amount of legislation can ever achieve the natural sense of respect and reverence for nature that is implicit in the culture of Native Americans and in other so-called primitive societies. A philosophy or science that insists on the finality of its own "correct" description of nature as reducible to mere inert matter is not only arrogant but immoral.

In recent years, Owen Barfield, the British philosopher and writer, has profoundly articulated the connection between morality and imagination—between what he calls " 'literalness' and a certain hardness of heart."

> Listen attentively to the response of a dull or literal mind to what insistently presents itself as allegory or symbol, and you may detect a certain irritation, a faint incipient aggressiveness in its refusal. Here I think is a deep-down moral gesture. . . . Instinctively he does not like it. He prefers to remain "literal." But of course he hardly knows that he prefers it, since self-knowledge is the very thing he is avoiding.[13]

Where today do we find such a "deep-down moral gesture"? Is it possible that despite all its brilliance and truth, physics has come to represent a certain hardness of heart, an immoral literalness? Once again, Barfield:

---

13. O. Barfield, *Saving the Appearances,* 162–63.

The relation between the mind and the heart of man is indeed a close and delicate one and any substantial cleft between the two is unhealthy and cannot long endure.

And finally, taking a cue from the ethical philosophy of William Blake, Barfield identifies the moral parameters of modern culture in no uncertain terms:

Imagination is the cardinal virtue, because the literalness which supports idolatry is the besetting sin of the age which is upon us.

CHAPTER 18

Particles and Unification

THE PERENNIAL SEARCH FOR THE elementary constituents of matter has plunged twentieth-century physics deep within the nucleus. There we have discovered a "zoo" of hundreds of exotic particles. A new theory was devised to make sense of this bewildering zoo. Two fundamental particle families—the *quarks* and the *leptons*—were proposed as the ultimate building blocks of matter. Everything else, from neutrons and protons to the most complex molecules, is made of quarks and leptons. These two particle families are so basic that they played a crucial role in creating matter in the earliest stages of the big bang. Thus, we see the rationale for the wedding of cosmology and elementary-particle physics.

In addition to the two basic particle families, there are four basic forces of nature—gravity, electromagnetism, and the strong and weak nuclear forces. In the latest theories, it is conjectured that these four forces evolved from one common unified force, which also had a starring role in the first formative moments of the big bang, some fifteen or twenty billion years ago. If these theories are correct, then the stalwart proton, which as far as we know has been around since the beginning of time, is unstable and will ultimately decay, like its many weak radioactive cousins. (So far, there's no evidence for the proton's instability.) Even quarks and leptons may be made of something yet more fundamental called *superstrings*. These are one-dimensional "strings" that are tightly coiled into inconceivably small particles. Furthermore, these tiny coils are conjectured to contain within them six (count 'em, six) as yet unknown and undisclosed dimensions of space and time. Even if these theories should prove wrong, they'll still command a high price in the science fiction market.

Particle Families

Our descent into matter has carried us from the world of everyday objects, through the realm of crystals and molecules, down to the level of the atom, and finally into the depths of the nucleus. In the effort to probe ever deeper into matter, physicists have penetrated even into the neutron and proton inside the nucleus. Not only have these nucleons revealed an internal structure of their own but they are also members of a bewildering family of particles, associated with the strong nuclear force. The particle members of this family are called *hadrons*, derived from the Greek word *hadros* (thick, bulky), which refers generally to the large size of these strongly interacting particles.

We have yet to decipher the complexity and variety of all the hadrons, but some of their characteristics fit into a pattern. There are two subgroups in the family—the *baryons* and the *mesons*. These two families differ in their relationship to the strong force. The baryons are the particles that "feel" the strong force. They interact, or attract and repel each other, by means of the strong force, just as charged particles interact by means of the electromagnetic force.

Mesons, on the other hand, play a role in transmitting the strong force between two baryons. They are the messengers of the strong force, so to speak. In quantum theory, particles do not mysteriously exert forces on each other across empty space. (This was the conceptual problem with the force of gravity that plagued Newton.) Instead, interacting particles transmit and exchange special messenger particles. This continual exchange causes the interacting particles to recoil and deflect as they emit and absorb the messengers. The exchange of messenger mesons between interacting baryons is the quantum mechanism of the strong force.

In the quantum theory of electromagnetism, by the same token, an electrical force is transmitted or carried between an electron and a proton by the so-called quantum of the electromagnetic field, which is the photon. As the electron and proton emit, exchange, and absorb photons between themselves, they rebound and recoil, just as a rifle and target do when they fire and intercept a bullet. It is this continual rebounding and recoiling of the electron and proton that adds up to their deflection and curving paths, which we take as the signature of the interaction between them.[1] There is no electrical force invisibly winging its way across space. Instead, the force is transmitted by the exchange of a photon or field quantum, which is a material particle.

In the case of the strong force, the mesons are the field quanta that transmit the force between interacting baryons. While in the case of the electromagnetic force, the mesons are the field quanta that transmit the force between interacting charged particles.

As it turns out, there is a family of particles associated with each of the four fundamental forces of nature—the strong, weak, electromagnetic, and gravitational forces:

---

1. Recall that forces that act at a distance, like gravity and electromagnetism, are identified and characterized by the motions and deflections they produce on particles.

Force/Field	Particle Family	Interacter Baryons (Fermions)	Messenger Mesons (Bosons)
Gravity	All matter	All particles	Graviton
Electro-magnetism	Charged particles	Electrons, protons, ions, alpha particles	Photon
Weak force	Leptons	Electrons, neutrinos, muons, tauons	W and Z bosons
Strong force	Hadrons	Quarks, nucleons, lambdas, sigmas, omegas	Mesons, gluons

The hadrons, as we have seen, are associated with the strong force. The hadron family has by far the greatest variety of different particles. The lepton family is associated with the weak force. It includes six interacting particles and three messenger particles, to which we shall return later. The family associated with the electromagnetic force has no special name or status as such. It includes all charged particles—electrons, protons, ions, and so on— as interacting particles and the photon as the sole messenger particle.[2] The family associated with the gravitational force is even more problematic. There is as yet no proper quantum theory of gravity, which is one of the greatest obstacles thwarting the efforts to unite or combine the four forces into one unified theory. Nevertheless, most physicists assume that a quantum theory of gravity will eventually be found. From the quantum point of view, this family must include all material particles as interacting members (because any two material particles attract each other gravitationally). In addition, there is a postulated messenger particle called the *graviton*, which has never been experimentally detected.

The interacting particles among the families are generally referred to as *fermions*. The baryons are fermions

---

2. Note that particles can be members of more than one family. Electrons are also leptons, and protons interact through the strong, the weak, and the electromagnetic force—not to mention gravity.

in the hadron family, while electrons and protons are fermions in the electromagnetic family. The messenger particles of all the families are called *bosons*. The mesons are the hadronic bosons, and the photon is the electromagnetic boson. (All of this technical vocabulary and language may seem excessively "botanical" and irrelevant to a conceptual understanding of physics. But because we are concerned with the unification of the four forces, and therefore of the four families, the names and classifications of the particles will prove helpful in describing the unification scheme.)

Thus, each force has its own family of fermions that interact through the exchange of bosons. This fundamental quantum explanation of force fields is called a quantum field theory. The granddaddy of all such theories is quantum electrodynamics—QED, for short—the quantum field theory of the electromagnetic force. QED is undoubtedly the most accurate and successful theory in the history of science. Its predictions have been experimentally verified, in some cases to twelve decimal places—one part in a trillion. It has become the model of choice for all the other forces. This model has been applied with some success to the weak and strong forces and with very little success to the gravitational force. The hope is that the quantum model will be common to all four forces and will provide the blueprint for their unification.

QUARKS

During the 1950s and 1960s, many new and exotic particles were found among the hadrons—the family of the strong force. In addition to the well-known proton and neutron— the two basic baryon interactors—heavier particles with strange properties were discovered, such as the lambda, the sigma, the xi (or cascade particle), and the omega. They were given Greek-letter names, with a superscript suffix (+, −, or 0) to indicate their electrical charge. For example, there is a positively charged sigma-plus, a negative sigma-minus, and a neutral sigma-zero. The pi meson, or pion, was also discovered as the boson, or messenger, for the strong force between the baryons. Other heavier mesons were found later—the K-meson and the eta-meson (usually called kaon and eta for short). Furthermore, all of these particles have antiparticles, and, with the exception of the neutron and proton, they are all unstable and decay into lighter particles.[3]

TYPICAL HADRONS

Fermions (Baryons)	Examples	Bosons (Mesons)	Examples
Nucleons	Neutron, proton	Pions	Π^+, Π^-, Π^0
Lambda	Λ^0	Kaons	K^-, K^0
Sigmas	Σ^+, Σ^-, Σ^0	Eta	η
Xi (cascades)	Ξ^-, Ξ^0		

This checkered family, containing several hundred members—old and new—in all their possible variations, was bewildering, to say the least. But patterns began to emerge. The new, more massive particles seemed to be "excited states" of the lighter ones. An excited state has higher energy and therefore, according to relativity, more mass than a ground state. For example, the lambda and sigma are excited nucleons, which decay to ground-state neutrons or protons, just as an excited hydrogen atom decays to its ground state. Similarly, the mesons decay to simpler, more stable states. But many of the newly discovered hadrons did not obey the same rules for decay as the known hadrons, and so the new ones were called *strange* particles. Furthermore, it was not clear what role all these new hadrons played in the scheme of things. The ordinary matter of the everyday world is made solely of neutrons and

---

3. Neutrons inside the nucleus are stable, but a free neutron is unstable and decays with a half-life of eleven seconds into a proton and an electron.

protons, so why were all these new baryons needed?

In 1964, Murray Gell-Mann and G. Zweig independently proposed a scheme for classifying the hadrons. Each hadron is assumed to be composed of a few simple entities, called *quarks*. Baryons, like the proton and sigma, consist of three quarks; mesons, like the pion and eta, consist of two quarks. To explain the hundreds of hadrons, eighteen quarks were introduced. The eighteen variations result from three different quark "colors" and six different quark "flavors." These terms have nothing to do with appearance and taste; they are whimsical names for certain esoteric properties of the quarks that relate to the nuclear force.

Color is the nuclear analogue of electrical charge. Here we see the characteristic effort to extend the QED model of electromagnetism to other forces—in this case, the strong force. Instead of the charge of the electromagnetic force, there are three colors associated with the strong force.[4] Color provides the attraction that holds the quarks together in the hadrons, just as electrical charge holds the nucleus and electrons together in the atom. Color is the "glue" of hadrons in the same sense that charge is the "glue" of atoms. Protons, neutrons, lambdas, and pions are indeed "hadronic atoms," and, at the next level, nuclei are "hadronic molecules." The electromagnetic force with its charge explains how electrons and nuclei combine into atoms and how atoms combine into molecules. Similarly, the strong force with its color explains how quarks combine into hadrons and how hadrons combine into nuclei. This is all a kind of hadronic "chemistry."

ELECTROMAGNETIC (CHARGE) CHEMISTRY

1 positive proton + 1 negative electron → 1 hydrogen atom
2 hydrogen atoms + 1 oxygen atom → 1 water molecule

HADRONIC (COLOR) "CHEMISTRY"

1 red quark + 1 green quark + 1 blue quark → 1 proton (hadronic "atom")
2 protons + 2 neutrons → 1 helium nucleus (hadronic "molecule")

---

4. More precisely, there are three colors and three anticolors, instead of the positive charge and negative anticharge of the electrical force.

A helium atom is in its ground state when its electrons have their lowest energies. When the electrons are somehow raised to higher energy states, the helium atom is in an excited state. Because higher energies correspond to higher masses ($E = mc^2$), an excited helium atom is slightly heavier than a ground-state helium atom. By the same token, three bound quarks with the least possible energy correspond to the ground state of a baryon, which is a proton or neutron. Nucleons—protons and neutrons—are the lightest baryons. Just as the normal helium atom represents the ground state of its electrons, so a proton represents the ground state of its quarks.

Lambdas and kaons represent excited states of a hadron, which decay, or "drop" down, to their lighter ground states, just as excited atoms do. The ground state of the lambda is the lighter neutron, and the ground state of the kaon is the lighter pion. But all excited states are fleeting and ephemeral: they decay. Molecules normally consist of atoms in their ground state, rather than their excited state. Similarly, nuclei are made of ground-state neutrons and protons. Nuclei—quark molecules—are made of nucleons, and nucleons—quark atoms—are made of quarks. An alpha particle is a quark molecule consisting of four quark atoms—two neutrons and two protons.

Nucleons and other hadrons can interact not only through the strong force but also through the weak force. A neutron, for example, decays into a proton, an electron, and a neutrino.[5] This is the prototype of the beta decay reactions that were discussed in Chapter 15. They are all governed by the weak nuclear force. So hadrons must also carry another kind of charge that is associated with the weak force. This is where quark "flavor" comes into the picture. Flavor is the weak-force analogue of electrical charge, just as color is the strong-force analogue. In fact, the six quark flavors correspond to the six members of the

5. Neutrinos are leptons and play an important role in the weak force, as we shall see.

lepton family—the family of the weak force. This symmetry between the two forces—six strong quark flavors and six weak leptons—is an example of the kind of link that physicists hope will lead them ultimately to the unification of the four forces.

By the way and for the record, the six quark flavors are called up, down, charm, strangeness, bottom, and top. The last two are sometimes called beauty and truth.

Quarks have other unusual properties. Because quarks make up a proton or neutron, quarks cannot have the standard quantized units of electrical charge. The up quark has a positive charge equal to two-thirds of a proton charge, and the down quark has a negative charge of one-third. A proton contains two up quarks and one down quark, for a total charge of one ($\frac{2}{3} + \frac{2}{3} - \frac{1}{3} = 1$). A neutron contains one up quark and two down quarks, for a total charge of zero. The arithmetic is right, but the properties are strange. Quarks are the only particles with fractional charges, i.e., charges smaller that the smallest quantized unit charge of a proton or electron.

When the quark theory was first proposed, experimental physicists quickly began looking for fractional charges in nature, which would provide indirect, but compelling, evidence of the existence of quarks. No fractional charges were ever found. Later it was decided on purely theoretical grounds that quarks must be permanently confined within their hadrons and can therefore never be directly detected in an experiment. Quark "confinement," as it is called, has become a fundamental tenet of the new quark theory. Despite the great explanatory power of quarks, it seems that we shall never have the pleasure of seeing one.

COLOR DYNAMISM

The theory of the strong force is modeled on QED—quantum electrodynamics. It is called QCD for quantum chromodynamics (or quantum color dynamics). The basic interacting particles, or fermions, are the quarks, and the basic messenger particles, or bosons, are the *gluons*. The attraction of colored quarks in QCD is explained by the exchange of gluons, just as the attraction of electrical charges in QED is explained by the exchange of photons. A neutron is a bound state of colored quarks that attract, just as a hydrogen atom is a bound state of charged particles that attract.

In Chapter 14, we saw how atoms bind together to form molecules. A molecule of hydrogen is the result of the covalent bonding of two uncharged hydrogen atoms. In covalent bonding, electrical charge is not the immediate "glue" of the bond. Quantum effects are primarily responsible for covalent bonding. Electrical charge plays only a secondary role. Thus, while the hydrogen *atom* is bound together as the direct result of the attraction of electrical charges (the proton and electron), the bonding of the hydrogen *molecule* is only indirectly due to electrostatic attraction.

The situation is analogous in QCD. The attraction of quarks via their color (color is hadronic "charge") is the bond that holds the proton and other hadrons together. Hadrons (quark atoms) are the direct result of the bonding of color. Nuclei (quark molecules) are bound together only indirectly by color (in analogy with the hydrogen molecule).

From the point of view of QCD, quarks interacting via gluon exchanges to form hadrons is the basic interaction. By contrast, baryons exchanging mesons to bond into nuclei is a derivative, or secondary, interaction. The baryons and mesons are the atoms of the strong force. The baryon-meson interaction that bonds nuclei is no more fundamental than the atom-atom covalent interaction that bonds molecules.

Quark color (which, as has been said, has nothing to do with the visible color we see) holds a proton together, just as electric charge holds a hydrogen atom together. But because the strong force is more complex than the electrical force, there are more colors than there are electrical charges. There are two kinds of electrical charges but six kinds of color (three colors and three anticolors).

With electrical charges, there is only one kind of attraction—between a positive charge and a negative charge (or anticharge). With color, attraction can result in two ways: a pi meson is the bound state of a quark and an antiquark, e.g., a red quark and an antired quark. A red and an antired quark produce a color neutral object—a "white" pion that has no color, just as a charged proton and an anticharged electron produce an uncharged object—an electrically neutral atom that has no charge.

But there's a second kind of attraction in QCD. A proton, for example, is made of three quarks having *complementary* colors—a red quark, a green quark, and a blue quark. Three complementary quarks attract. Red, green, and blue quarks make a "white" proton, which is color neutral.[6] A proton is colorless or color neutral, just as an atom is charge neutral.

Thus, there are two ways that quarks bond to form hadrons. A red, green, and blue quark exchange gluons to form all the color-neutral baryons, from the ground-state proton and neutron up to the heavy excited omega. A quark and an antiquark bond together to form all the color-neutral mesons—pions, kaons, and so on. Mesons follow almost exactly the pattern of atoms, with their equal amounts of positive charge and negative anticharge. Therefore, all hadrons are made of either three quarks of complementary colors (the baryon pattern) or a quark and an antiquark (the meson pattern).

---

6. The idea of complementary colors, which combine to make white (considered here to be colorless) is used in an analogy with opposite charges that neutralize each other. Quark color is not a visible or even directly detectable physical property. The names red, green, and blue are simply a whimsical choice of convenience.

It is still not clear why so many different kinds of hadrons exist. They are mostly excited states of quarks, some of which are pretty exotic. Everyday matter is made exclusively of neutrons and protons, which consist of only two quark flavors—up and down.

Ordinary matter contains only two kinds of leptons, as well—the electron and its neutrino. So why do we need charm, strangeness, top, and bottom at all? For that matter, why do hadrons have the particular masses and quantum properties that they do? Is there any pattern to it all?

The answer may have something to do with the level of symmetry required for the unification of the forces. Unification, however, is far from an established theory. For the moment, the complex variety of hadrons is still something of a puzzle, despite the simplification provided by their underlying quark structure.

Symmetry and Force

In Chapter 15, we noted an important characteristic feature of nuclear reactions. When a neutron undergoes beta decay to a proton, or when nitrogen is transmuted into oxygen, the total number of nucleons is preserved, as is the total amount of electrical charge. We say that these reactions are governed by two conservation laws—conservation of nucleon number and conservation of charge. The preservation or conservation of some quantity during a change or transformation is a fundamental feature not only of nuclear reactions but of all interactions among elementary particles. It applies therefore to all four forces of nature, which, after all, provide the basis of all interactions.

Conservation during change is also called *symmetry* in quantum physics, and the meaning of the term is somewhat different from its meaning in geometry (although the two meanings are related). In physics, symmetry means that some property is the same at different *times*. The number of nucleons is the same, or *symmetric*, before and after a nitrogen transformation. In geometry, we refer to something that is the same at different *places*. A Rorschach pattern looks symmetric (only reversed) on the left and right. Symmetry in physics also refers generally to a preserved quantity rather than to a preserved shape.

As an analogous example, consider a transformation of a simple geometrical figure. Suppose a five-pointed star can rotate about an axis through its center. In this analogy, we shall be looking for some conservation in the appearance of the star under a rotational "transformation." Let's suppose the star initially has it uppermost point exactly in a vertical position. We designate this particular orientation as the appearance, or "state," of the star that we wish to preserve.

As the star rotates, its state continually changes—a snapshot taken at any instant will in general not line up exactly with the original orientation of the star. But at certain points in the rotation—at one-fifth of a circle, two-fifths, and so on—the star's position, or state, will exactly overlap the original orientation. If there are no distinguishing marks on the star, an observer will not be able to tell the difference between the original and the rotated state. So for an arbitrary transformation—a rotation by an arbitrary amount—the original state of the star is not conserved. But for rotations of one-fifth of a circle, two-fifths, three-fifths, four-fifths, or five-fifths, the original state is conserved. We say that this star has a five-part symmetry, corresponding to rotations by fifths of a circle (or multiples of 72°).

In the field theories for the forces, symmetry plays an essential role. In analogy with the five-part symmetry of the star, QCD has a three-part symmetry associated with the color force, because there are three colors and three anticolors. QED has a one-part symmetry associated with the one kind of electrical charge and its anticharge. QCD is said to be symmetric because the interaction is unaffected by a transformation of the colors into one another.

Changing red into green, green into blue, and blue into red in a proton, for example, still leaves the proton white or colorless. It is just like moving the star through 72° or 144°, and so on. Nothing really changes. The strong force interaction is conserved under color transformations, just as the star's state is conserved under rotations through 72°.

This idea of symmetry and conservation is the unifying theme behind the quantum field theory for each of the forces of nature—gravity, electromagnetism, and the strong and weak forces. By searching for symmetry principles that are common to the four forces, physicists hope to discover a field theory that is general enough to describe all interactions and to demonstrate as well that the four separate forces have evolved from one common unified force or field.

We know, for example, that QED has one-part symmetry and QCD has three-part symmetry. Perhaps a unified theory with eight-part or twenty-part symmetry will be capable of explaining all the forces and particles and their interactions. This is a tall order and far from complete, but it has become the guiding principle for work in elementary-particle physics and cosmology during the last decades of the twentieth century.

We can understand how forces come about from the point of view of symmetry by returning to our rotating star example. Forces are the result of a breakdown or reduction of symmetry. If one of the five points of the star is broken off, then the star's symmetry is destroyed or broken. No rotation (except through 360°) will now leave the state or original appearance of the star unchanged. The symmetry has been broken (or at least reduced from five-part to one-part symmetry). To restore or correct the broken symmetry requires the action of some external agent or force—something that will reinstate the missing point. A force remedies the broken symmetry. In field theory, a force is always tantamount to the restoration of a broken symmetry.

Inside of a proton, the red, green, and blue quarks interact by exchanging gluons. For example, a red quark interacts with a green quark by exchanging a red-green gluon. It turns out that when the red quark emits the red-green gluon, it immediately turns green. Momentarily, there are two green quarks and a blue quark. There is a color imbalance. The proton is no longer color symmetric or invariant to color transformations among its quarks. The symmetry of three complementary colors has been broken. To restore the red-green-blue symmetry, the green quark must turn red. The exchanged red-green gluon is the agent that transforms the green quark into a red one. The gluon restores the broken symmetry.

The exchange of gluons between quarks is the mechanism by which the color force operates in QCD. Quarks attract and bond in the proton by exchanging gluons, just as the electron and proton in the hydrogen atom bond by exchanging photons. This boson (gluon or photon) exchange mechanism is the explanation for forces in all quantum field theories. But boson exchange, as we have just seen, is also the vehicle for breaking and restoring symmetry. Quarks attract each other in the proton by exchanging color-transforming gluons among themselves. This exchange also breaks and restores color symmetry. We can describe forces in quantum field theory either as an exchange of bosons between fermions or alternatively as the breaking and restoring of a symmetry. The latter description, although more abstract, is the one that theoretical physicists believe holds the greatest promise for the construction of a unified theory of all the forces.

The Weak Interaction

The strong, weak, and electromagnetic fields are each described by a quantum theory in which a boson exchange between fermions is the mechanism of attractive and repulsive forces. It is also the mechanism for restoring broken symmetries. In QED, only one boson—the photon—is exchanged, because the charge of the electron is not changed (or rather is only "transformed" into itself).[7] In QCD, quarks of three different colors may be transformed into each other, which requires eight gluons.[8] But there is one more force we must now discuss—the weak interaction.

The lepton family interacts via the weak force. There are six members of the family: the electron, the muon, the tauon, plus an associated *neutrino* for each—an electron neutrino, a muon neutrino, and a tauon neutrino.

The neutrinos are among the queerest animals in the elementary-particle zoo. They were originally predicted in the 1930s to account for the apparent failure of the laws of conservation of energy and momentum in beta-decay reactions. When a neutron decays into a proton and an electron, momentum and energy are apparently lost. Wolfgang Pauli and Enrico Fermi proposed the existence of a new particle—the neutrino—that accompanies the proton and electron and carries off the missing energy and momentum.

Neutrinos have no mass and no charge, but they do "spin" and travel at the speed of light.[9] You may wonder why a neutrino is considered a particle and how it can be detected. Well, there are other massless particles—photons, for example. As for detection, it ain't easy. Neutrinos interact so weakly with other particles that they pass through the earth and most of the matter in the universe with hardly a notice. In fact, it took twenty years for neutrinos to be detected after their prediction. They don't call it the weak interaction for nothing.

-----

7. The one-part symmetry of QED does not give us much insight into the electromagnetic force. It is in relating the different forces that the one-part QED symmetry becomes important, because of its formal connection with the multiple symmetry of the weak and color forces. In fact, there is an undetectable property of electrons that changes in photon exchanges. It is called the *phase* of the electron wave, but we shall bypass this subtlety.

8. There are really nine gluons for the 3×3 possible transformations, but one of them is mathematically redundant.

9. There are theoretical and cosmological arguments for allowing the neutrino to have a very small mass, but experimental evidence is still consistent with a zero mass.

To complicate matters even more, we now recognize the existence of three different neutrinos, one associated with each of the other leptons—the electron, muon, and tauon. As with hadrons, we're not at all sure why so many leptons exist, because only the electron and its neutrino show up in ordinary garden-variety matter.

The six leptons correspond to the six quark flavors, suggesting a parallel symmetry between the two families. Despite the close apparent relation between hadrons and leptons, the first real victory in the quest for a unified field theory was the unification of the weak force with the electromagnetic force rather than with the strong force.

THE ELECTROWEAK UNIFICATION

In the simplest and most basic form of the weak interaction, an electron may exchange a boson with another electron or with an electron neutrino. This requires a two-part symmetry because an electron may become an electron (an identity transformation) or an electron neutrino.[10]

This is the typical pattern in quantum field theory. The exchange of a boson between two fermions accounts for the force. But the emission and absorption of the boson also change some characteristic property of the fermion. A quark changes color when it emits a gluon—a red quark becomes green or blue. (A red quark can also "become" a red quark in an identity transformation.) Similarly, when a lepton emits a boson, it changes its flavor: an electron can become a neutrino or an electron. In the case of the strong force, transformations among three colors implies a three-part symmetry, requiring nine gluons.[11] Thus, for the simplest weak interaction, transformations between two flavors—electron and neutrino—implies a two-part symmetry, requiring 2×2 or four bosons.

Now electrons can also interact through the electro-

---

10. For convenience, electron neutrinos will be referred to simply as neutrinos in this section.
11. See footnote 8.

magnetic force—that's QED. So the interaction between two electrons is really a combination of the electromagnetic and weak forces. This is the clue to combining or unifying these two forces. In 1967, Sheldon Glashow, Abdus Salam, and Steven Weinberg proposed a theory that unified the electromagnetic and weak forces.[12] The new theory is called *electroweak*, or simply EW for short. EW combines the four bosons of the weak interaction with the one photon of QED into four new EW bosons for the four possible EW transformations: electron to electron, electron to neutrino, neutrino to electron, and neutrino to neutrino.

Under the normal present-day conditions of matter, the electroweak unification is broken apart into the separate weak and electromagnetic forces that we observe. But if we can artificially create the conditions for the electroweak unification, then we should discover the new bosons predicted by the EW theory. The predicted bosons were experimentally found in 1983 at the CERN Laboratory in Switzerland in a heroic team effort led by Carlo Rubbia, who received the Nobel prize for this work the very next year. The four EW messenger particles, or bosons—the photon, the W^+, the W^-, and the Z^0—mediate the four EW transformations between electrons and neutrinos. This new theory represents the first effort, and so far the most successful one, at unifying the four forces.

SYMMETRY, UNIFICATION, AND COSMOLOGY

If the EW theory is valid, and if an ultimate unified theory is a real possibility, then why do we see separate rather than unified forces today? The answer has to do with the evolution of the universe and the corresponding breaking of symmetry.

Earlier we discussed the five-part symmetry of a rotating star. The star's original orientation, or state, is con-

---

12. Glashow, Salam, and Weinberg received the Nobel prize for this work in 1979.

served under rotations that are multiples of 72°. We've also discussed the three-part, two-part, and one-part symmetries, respectively, of the color force, the EW theory, and electromagnetism. We can easily imagine higher level symmetries by using stars of six points, or ten, or a hundred, or a thousand. If we use polygons instead of stars, then we see the same symmetry progression with squares, pentagons, octagons, and so on. As we progress to polygons with more and more sides, we approach the limiting case of a circle, which can be rotated about its center through any angle and still preserve its original appearance, or state. A circle may be said to have an infinite degree of symmetry, or continuous symmetry, or even "perfect symmetry."

The unification of the four forces is based on the assumption that the cosmos in the very earliest stages of the big bang had perfect symmetry, characterized by the one unified force. As the universe expanded and cooled, the perfect symmetry was broken, resulting in two, three, and finally four separate forces. The breaking of symmetry as the universe cools is a very difficult concept to understand and even to describe precisely in the proposed unification schemes, all of which are quite hypothetical. We can get an inkling of it, however, by using further analogies.

The circle in the example we have discussed has a higher degree of symmetry (infinite) than an octagon (eight), and an octagon has a higher degree of symmetry than a pentagon (five). Breaking a symmetry pattern means reducing its degree of symmetry. When we break off one point of a star, we reduce its symmetry from five-part to one-part symmetry.

Let's apply this logic to a different, more physical, case. When we freeze water into ice, we reduce its degree of symmetry. Water molecules in the liquid state are in a random disordered pattern. But in ice, the molecules form a regular crystal lattice. The random pattern of molecules in liquid water would look the same no matter how a sample of water is rotated. Therefore, water in its liquid state has an infinite degree of symmetry like a circle. But the ordered molecules in ice have a finite degree of symmetry, corresponding to the restricted number of ways that the rotated crystal would look like its original orientation.[13] In this sense, we may say that freezing the water reduces, or breaks, its symmetry. By the same token, as the universe cooled, the symmetry of its unified force was broken, or "frozen" out.

The precise mechanism for the breaking of cosmic symmetry with falling temperature is still unclear and conjectural. A separate new field—the *Higgs field*—has been proposed to explain this process. It remains, however, quite hypothetical. The search for the boson of the Higgs field is one of the important experimental problems proposed for accelerators of the future.[14]

Nevertheless, even if we don't know exactly how or why symmetry breaks, we do know when it's supposed to have happened. Perfect symmetry and the one unified force presumably prevailed during the first flickering instant after the cosmos was born. To be precise, it lasted about one ten-millionth of a trillionth of a trillionth of a trillionth of a second. That's a decimal point followed by forty-two zeros and a one. Such an infinitesimal interval of time is vastly beyond our measurement capabilities, but there are quantum theoretical justifications for such a small number. What's more, we are contemplating here not only an impossibly minuscule instant of time but such an instant that's supposed to have occurred fifteen billion years ago. This may seem like a clear example of the folly of the quantitative, but many physicists and cosmologists take it quite seriously.

The initial "era" of unification and perfect symmetry ends as the cosmic temperature drops to a cool million

---

13. Recall that the term *symmetry* in physics does not have its usual geometrical usage. Geometrically, ice crystals would seem to be more symmetrical than amorphous water. But the reverse is true in the quantum-theory sense of symmetry.

14. See M. J. G. Veltman, "The Higgs Boson," *Scientific American* (November 1986): 76.

trillion trillion trillion degrees Celsius. In the second era, there are two separate forces—gravity and the still unified strong, weak, and electromagnetic force. This latter union is called the *grand unification*, and this era is named after it. The era of grand unification lasts until the cosmos is a ten-billionth of a trillionth of a trillionth (a decimal point followed by thirty-three zeros and a one) of a second old. Further cooling and symmetry breaking brings the cosmos into the electroweak era, characterized by three separate forces—gravity, strong, and electroweak. This era lasts until the universe is all of one-trillionth of a second old. At this point in time, the cosmic temperature was a mere quadrillion degrees, and the four forces we know and love so well froze out and have lasted ever since. Thus, any unification of the four forces ended within the first trillionth of a second of the life of the cosmos. Since then, it's all been smooth sailing (but for the creation of atoms, stars, galaxies, and . . . oh yes, us).

This scenario of the cooling of the universe and the parallel freezing out of the forces of nature is unique in the history of science not only for the remarkable story it tells but also for the unheard of marriage of quantum theory and cosmology. The macro- and microrealms of the physical universe, to everyone's surprise, have been combined into one interdependent picture of the universe. It remains to be seen how accurate this picture is, whether it can withstand the tests of future observations and experiments, and whether it can successfully be formulated into a complete mathematical "theory of everything." But it has become the prevailing, challenging, and inspiring view of a new fin-de-siècle physics.

A GUTSY THEORY

So far in this chapter, we've looked at how the four natural forces are related to the elementary particles and their families. We've seen how the basic quantum-force mechanism—an exchange of bosons between fermions—is used to explain the electromagnetic, strong, and weak interactions (although not yet the gravitational interaction). The central features of this picture are combined into the "standard model," in which the three nongravitational forces and their particle families are explained in terms of eighteen quarks, six leptons, and their interactions. According to the standard model, the whole world is made out of just quarks and leptons.

But the standard model is really like three separate theories (or two if EW is completely correct), with gravity standing apart as yet a fourth theory. This situation is aesthetically unsatisfying to most physicists. They hope for a much simpler alternative—a single encompassing theory to explain both the evolution and the behavior of matter. So there is a quest today for a unified theory of all the forces and particles—a quantum-field "theory of everything," based on the idea of broken symmetry.

We've seen, too, how the idea of broken symmetry is applied to forces and their unification. So far, we've explored the symmetry approach in the unification of the electromagnetic and weak forces and in the overall scenario for the cosmic evolution of the forces. What remains is to look at the proposed schemes for the unification of all the forces—first combining the electroweak and strong interactions and then adding gravity for the final unification.

The theories that attempt to unify the strong and EW (electroweak) interactions are called grand-unification theories, or GUT. In such a theory, the electromagnetic, the weak, and the strong forces are described in terms of a single unified interaction. Since electromagnetism and the weak force are already united in the EW theory, what is needed is a combination of the electroweak and color forces. The simplest scheme for this enlarged combination is a theory with five-part symmetry, corresponding to three quark colors and the first two lepton flavors. It is a theory in which five fermions—a red, green, and blue quark, an electron, and a neutrino—attract and repel each other by the exchange of bosons. As in the simpler cases of QED,

QCD, and EW, a boson exchange momentarily upsets and then restores a symmetry balance. In QCD, for example, a blue quark emits a gluon and turns green. The gluon is absorbed by a green quark that turns blue. A force acts between the quarks, the blue-green symmetry imbalance is restored, and each quark undergoes a change of color.

In the QCD exchange, the quarks (fermions) change color. In the EW exchange, the leptons (fermions) change flavor. In the enlarged exchange of a GUT interaction, the fermions (quarks, electrons, neutrinos) change color or flavor. As we move to higher levels of unification, more dramatic fermion transformations become possible. In QCD, a quark can change its color. In EW, an electron can change into a neutrino. In GUT, the same color and flavor changes are possible. But because there are five interacting particles—three quarks and two leptons—a green quark can now change into an electron or a neutrino can change into a red quark. Transformations between leptons and quarks are now possible.

If one of the three quarks inside a proton changes into an electron, which flies off, that leaves a two-quark pion behind. This leads to the remarkable and revolutionary prediction of the instability of the proton. A proton can decay into simpler particles—a pion and an electron. Nucleon number is no longer conserved in such reactions. Instead, the higher symmetry of the GUT interaction introduces an enlarged conservation law in which leptons and quarks are conserved together rather than separately.

This incredible prediction set off a worldwide race among experimental physicists to find the first proton decay. Naturally, the decay rate had to be very, very low, for otherwise normal matter, made largely of protons, would be highly unstable. There's good reason to believe that the protons we "see" all around us in the cosmos have been alive and kicking for fifteen billion years.

The original GUT predicted a proton half-life of a million trillion trillion years (a one with thirty zeros after it). That's one hundred million trillion times the present fifteen-billion-year age of the universe. Few physicists would be willing to wait that long for a proton to decay. But since we're dealing with statistical averages in decay processes, it's much easier to watch a million trillion trillion protons for one year. Among that many protons, there's a fifty-fifty chance that one will decay in a year's time.

In practice, several different experimental groups amassed about ten tons of matter, which contain enough protons to produce one or two decays a month, according to the GUT prediction. The material can be almost anything because all matter contains about the same number of protons by weight. Water, concrete, and steel were used in different experiments. To guard against mistaking some rare spurious event for an actual proton decay, the experiments were mounted far below the earth's surface to eliminate all but the most powerful cosmic rays. One was done in a tunnel under the Alps and another in a deep abandoned iron mine in northern Minnesota. Even at these depths, spurious subatomic processes will occur, and so extraordinary care is required in the electronic detection and analysis of all events.

A number of experimental groups watched for several years and never saw a single proton decay.[15] That meant one of two things: either the proton half-life is longer than predicted or the theory is completely wrong. New experiments were later mounted with about a thousand tons of matter, and still no proton decays have been seen. No definitive conclusion can be reached. An experimenter can only place a lower limit on the lifetime of the proton, which is currently about 250 times longer than the original prediction.

New GUTs are already in the works with more complex mechanisms that allow for longer proton lifetimes, but such lifetimes are beginning to exceed the capacities of current experimental methods. The new GUTs will re-

15. See J. M. LoSecco, F. Reines, and D. Sinclair, "The Search for Proton Decay," *Scientific American* (June 1985): 54.

quire different and even more challenging experiments for their verification.

THE PROBLEM OF VERIFICATION

The problem of measuring the proton half-life is only the tip of the iceberg. The general verification of the new unified field theories is far beyond the capacity of current experimental technology. The proton decay experiments, as difficult as they are, can at least be carried out under the normal conditions that prevail in ordinary matter. Other experiments to verify GUTs will require far more extreme conditions. This is because of the exceedingly high temperatures at which the various stages of unification occur. As we've already seen, the lowest unification—the electroweak—occurs at a temperature of a quadrillion degrees. The equivalent of such temperatures can be achieved today only in a particle accelerator. Indeed, the discovery of the EW bosons at the CERN accelerator was done very close to the limit of the energies attainable today.

The temperature of an object is a measure of the average speed of its constituent atoms, and the speed of an atom is directly related to its energy. In a hot body, the atoms have more energy than at room temperature. The biggest accelerators today are capable of reaching particle energies of a trillion electron volts.[16] This is equivalent to a temperature of a quadrillion degrees—the temperature at which the EW unification sets in. Thus, it is possible to perform experiments at the lower edge of the range of EW unification. But grand unification occurs at temperatures and energies a trillion times higher—far beyond the wildest projections for future accelerators. The proposed SSC—the superconducting supercollider—will increase the highest energies of today's accelerators by a mere factor of twenty. No known techniques in the foreseeable future can attain the required GUT energies. Furthermore, the temperatures

---

16. This means that an electron or proton is accelerated through a voltage of a trillion volts.

and energies at which all four forces are unified is another factor of a million higher than the GUT values.

One of the biggest objections to current theoretical speculations on unification is the almost certain impossibility of any timely experimental verification. It will be a herculean task for both theoretical and experimental physicists to find some consequences of the new theories in a low enough energy range to be observed. Physicists can be remarkably resourceful. Will they be able to overcome the insurmountable energy barriers of unification?

SUPERSYMMETRY AND SUPERSTRINGS

The ultimate goal of the unification program is to combine all four forces of nature into one unified field. Such a theory would unify GUT with gravity and reach the highest possible level of symmetry; such theories are referred to as *supersymmetry* theories, or SUSYs. The symmetry of a SUSY theory implies even more remarkable particle transformations than in EW and GUT. In EW, electrons can become neutrinos; in GUT, quarks can turn into electrons. But in SUSY, a fermion can become a boson—a quark can become a photon, or a neutrino can turn into a gluon. In other words, the world is so symmetrical in SUSY, that there is no distinction between the interacting and messenger particles. The particles that sense the force and the particles that transmit the force are interchangeable. In effect, there is no meaningful difference between the particles and the forces. They are one and the same. (Talk about symmetry!)

But the dream in this case is far from the reality. In addition to the burdensome experimental difficulties we've already mentioned, there are also two horrendous theoretical problems. One is the construction of a quantum theory of gravity, which would be required for any unification with GUT. But general relativity—the only known and accepted theory of gravity—is a geometrical theory, not a quantum-field theory.

Gravity is the result of large-scale space-time curvature, and not of any microscopic boson exchange. Even the gravitational waves, predicted by general relativity, have so far escaped detection, let alone the quanta or bosons of these waves—the elusive gravitons—that a quantized theory of gravity must have. The mathematics of quantum gravity has also turned out to be all but intractable. Infinite quantities appear that cannot be eliminated, and there is no clear method for calculating the physical effects of the theory—no conceivable experimental test as yet.

The other problem is the proliferation of the new particles needed for the enlarged symmetry of the unified SUSY force. EW has four bosons, QCD eight, and GUT twenty-four in its simplest formulation, which is probably incorrect. Where SUSY will end up is anybody's guess. Hundreds of new particles have already been conjectured, and even if they make more sense than the old hadron zoo, none of these new particles has any remote possibility of being detected in the foreseeable future. The prospects for a realistic SUSY are very dim. The shotgun wedding of relativity and quantum theory may be annulled before it takes place.

But there is one glimmer of hope on the horizon—superstrings. In all previous theories, the elementary constituents of matter are assumed to be particles—tiny point-like objects of neglegible or zero size. Now points have zero dimensions in geometry. Regions of space are three-dimensional, surface areas are two-dimensional, curves are one-dimensional, and points have no dimensions. The most basic elementary particles—quarks and leptons—have no detectable size. They are treated as zero-dimensional in the standard model.

By contrast, the theory of superstrings assumes that the elementary objects are one-dimensional, like a string. But they are strings coiled up into such tiny balls that they appear to be points. Why, you may inquire, would anyone want to make such an assumption? The answer is that in superstring theory it is possible to describe all the particles as various vibrational states of the superstrings. When the string vibrates one way, it looks like a neutrino; another way, like a quark; and yet another way, like a proton. So this promises a possible rationale for all particles.

Even more tantalizing is the fact that not only particles but even the properties of space-time itself seem to arise from superstring vibrations. Both the general theory of relativity and quantum theory have some chance of being pulled out of the superstring hat, provided that we can deal with the proliferation of dimensions required by superstring theory.

Superstrings, as it turns out, need more than one dimension: they need ten. To accommodate the high level of symmetry inherent in such an all-encompassing theory demands a space of more than three dimensions. It requires ten dimensions—nine for space and one for time. So the four-dimensional space-time of relativity is included, but where are the other six dimensions? It turns out they're coiled up inside those superstring balls.

Imagine rolling up a two-dimensional rectangular sheet of rubber into a tube. The tube still has two dimensions. Now stretch the tube out, making it longer and thinner as you go. The rubber, of course, is ideal—perfectly flexible. You can stretch it so thin that it looks like a line with only one dimension. It's so thin that you can no longer detect its second dimension. Now coil the long thin tube up into a tiny ball that is so small that it appears to have no size. It looks like a zero-dimensional point. This is the recipe for reducing two dimensions to none at all.

In superstring theory, the original ten dimensions of the primordial cosmos were somehow or other transformed into six-dimensional superstrings and four-dimensional space-time. This may have happened with the cooling and symmetry breaking of the cosmos, as in the "old" unified-field theories. How this happened, why the six dimensions were "compressed," what manifestations these hidden dimensions may have, and how they may be experimentally detected are all open questions that superstring theorists must ultimately deal with. Let's hope they don't get all tied up in knots.

CHAPTER 19

Physics and Mythology

CONJECTURAL THEORIES, LIKE THOSE OF unification and superstrings, often suggest images of fantasy and science fiction. But in a subtler way, many scientific concepts and theories have psychological, mythical, and even spiritual significance, of which we are largely unconscious. Sigmund Freud, Norman O. Brown, and Carl Jung, among others, have emphasized the fundamental role that instincts, fears, and archetypes play in our psyches and activities. Our grand scientific theories and masterful technology may be psychological schemes to create the illusion of our heroic stature, to insure our scientific immortality, to proclaim our cultural superiority, or to demonstrate our intellectual prowess. To what extent have we created scientific concepts and theories to satisfy our deepest psychic and spiritual needs?

It is fruitful to explore physics from a psychic and mythical point of view and to compare its metaphors with those of alternative systems of knowledge, belief, and wisdom. In many primal and ancient cultures, matter is not inanimate but has spirit or soul. Space is not empty but enfolds within it intelligence and insight. Time is not linear and relentless but has a cyclic, cumulative, and enriching quality. To our peril, we have allowed science to discount and reject this precious lore. Yet science has its own nourishing mythology and spirit, if only we are willing to acknowledge it.

MYTHOLOGIZING PHYSICS

It may seem incongruous and anachronistic to look at physics from a mythical point of view. Scientists pride themselves on the achievements of the scientific era and attribute their success to the rejection of myth, superstition, and anthropomorphism. But it has become clear in the closing years of the twentieth century that both the theoretical ideas and the practical consequences of science are capable of doing harm as well as good. Quantum theory provides no clear or meaningful picture of a "physical world" but only abstract probabilistic calculations. The legacy of technology includes the bomb and pollution.

In exploring physics from a humanistic point of view, which is the program of this book, it is necessary to take subjective, aesthetic, and psychological considerations into account. Physics simply is not an objective body of facts and established theories. It incorporates its own system of beliefs, conventions, tacit knowledge, biases, assumptions, misconceptions, lore, and myth. It cannot stand above aesthetic judgment and psychological analysis. While Freud's couch isn't as mandatory or authoritative as it once was, the idea that there are unconscious motives and themes in all human endeavors is here to stay.

The desire for power and the fear of death undoubtedly play a role in physics. Sigmund Freud, Brigid Brophy, and Norman O. Brown have emphasized how deeply our psyches and lives are influenced by our knowledge and fear of death.[1] No one has explored better than Ernest Becker

---

1. S. Freud, *Beyond the Pleasure Principle* (London: Hogarth Press, 1961); B. Brophy, *Black Ship to Hell* (New York: Harcourt, Brace and

how the denial of death has profoundly affected all human activity.[2] In another book, I have tried to apply Becker's analysis to physics.[3] I believe that the most basic concepts in physics—space, time, matter, and number—have a metaphorical character, which, among other things, unconsciously denies death.

Space, for example, is the very medium of existence itself—a guarantee against oblivion and the loss of being. The word *exist* comes from Latin and means to stand out. Space is precisely what we stand out from. It is the background against which we contrast and articulate our being, our individuality, our ego, our existent selves. Physics, in fact, has crystallized and promoted just this meaning of space. The empty void of physical space is the all-encompassing container of things, matter, bodies, beings. It is the platform, the scaffolding, which supports matter and provides it with place, location, extension, room—all the things we take for granted as individual, articulate beings. To treat space as the mere abstract continuum of mathematical physics is to deny its very real metaphorical, psychological, and mythical character.

The physical concept of energy has its mythical aspect in the form of heat, activity, volatility, dynamism, animation, spirit, and life force. All of these have mythological, allegorical, and divine counterparts: Agni, the fire god of the Vedas; Dionysus, the Greek god who could incite madness and frenzy in humans; Mars, the dynamic Roman god of war; *prana*, the inspiriting life-giving power of the Hindu cosmos; and *rajas*, the animating activating element of the *Bhagavad Gita*, that together with *satva* (being) and *tamas* (inertia) forms the trilogy of elements that compose all things.

The concepts of physics—space, time, matter, energy,

World, 1962); N. O. Brown, *Life against Death* (New York: Random House, 1959).

2. E. Becker, *The Denial of Death* (New York: Macmillan, 1973).

3. R. S. Jones, *Physics as Metaphor* (Minneapolis: University of Minnesota Press, 1990), Chapter 7.

charge, temperature, color, charm, flavor, spin, wave, particle—are all so pregnant with mythical and metaphorical meaning that they can just as well be viewed as Jungian archetypes.

MYTHIC THEMES

It is not only the mythical content of the ideas and concepts of physics but the mythical force of many of its assumptions, laws, and theories that unconsciously influences us. We have already discussed the big bang as a creation myth. Scientists may balk at applying a mythical status to a scientific theory, but it is not entirely inappropriate. To begin with, there are conjectural beliefs and assumptions within the big bang theory that are incapable of verification and that will always remain articles of faith. Have the laws of physics been unchanging over fifteen billion years? Have space and time always been the same? What does it mean to describe a state of the universe that never has been observed and never can be? Where did all the matter, energy, and space-time of the universe come from? What initiated the big bang?

But more to the point is the psychological force and unconscious influence of the big bang, which is taken as an authoritative description of how we got here and why. The big bang scenario implies a universe devoid of meaning, purpose, and value. Because science is so widely and naively accepted and because it effectively plays the role of a state religion in our society, the big bang has become our consensus view of reality—our creation myth—that counters and rejects all previous religious and mystical views of existence.

Other candidates in physics for themes with powerful mythical force are the conservation of energy (in fact, the whole idea of conservation laws) and the materialistic and reductionist presuppositions of modern science. We have inherited from the Greeks our quest for permanence and constancy amidst the flux and change of the material world. There is no reason whatsoever to believe in the ultimate truth of such an idea. The 2,500-year-old religion of the Buddha is based on quite the opposite belief, namely, that in its deepest nature all of existence is fleeting and impermanent. Nothing is conserved.

As for materialism and reductionism, they are canons of faith, and prejudiced ones at that. We haven't a particle of evidence that consciousness, life, and history can be reduced to the dance of electrons. We can't even derive the wetness of water from the properties of the hydrogen and oxygen atom. Are matter and spirit separate but equal, the same but different, illusions, reality? Science may assume answers to these questions but can prove nothing. It's all a system of belief—mythology.

Then there are physics myths that characterize the twentieth century, but they are really nothing new. Consider, for example, relativity, quantization, complementarity, indeterminacy. That different observers should describe reality differently would hardly surprise mystics, seers, and practitioners of meditation. A quantized discrete structure of reality is part and parcel of *I Ching* divination, astrological analysis, and cabalistic mysticism. Complementarity is fundamental to Taoist philosophy, as Bohr knew only too well. He incorporated the Taoist yin-yang symbol into his personal coat of arms. As for indeterminacy, the arbitrary and whimsical interference of the Greek gods in natural events is an easy forerunner of quantum randomness.

Science plays a central role in the human desire and need to rationalize and make sense of existence. Even the denial of sense has a logic and mythical character of its own. Whether it seems appropriate or not, we cannot deny the psychological, religious, spiritual, and mythical force of science.

Promethean Fire

We can use a mythological analysis to help us make sense of and interpret science. As an example, we shall explore our efforts to harness the energy of fusion and the sun in terms of the myths of Frankenstein and Prometheus.

The Frankenstein myth, created by Mary Wollstone-craft Shelley, has long been considered an allegory of the evils of science. Its popular image is one of the intense scientist gone mad, but the story is more complex and subtle than that. It reflects some of our deepest fears and unconscious beliefs concerning the destructive potential of science. Shelley consciously modeled her tale on the Greek myth of Prometheus, the god who created humanity from clay and water. Dr. Frankenstein represents the dark side of Prometheus. To fashion his creature, Frankenstein does not use fresh clay from the earth as in the Greek myth but morbid decaying flesh that had been obtained from the soil of a graveyard and from dissecting tables. It is not wise Athena who breathes life into the monster; violent Zeus energizes him with a lightning bolt. Furthermore, Dr. Frankenstein provides his monster with no knowledge, skill, or wisdom. It is a blind, rash creation of an accidental and meaningless life.

In contrast with Frankenstein's dark ambitions and hunger for power, the motives of Prometheus are noble and altruistic. Prometheus does not merely create human beings as blank slates. He bestows upon them the divine gifts of architecture, astronomy, mathematics, navigation, medicine, metallurgy—the arts and crafts that add meaning to life. These skills Prometheus learned from Athena, who was born from the head of Zeus and is herself the goddess of wisdom. This creation of flesh together with intelligence and meaning symbolizes the unity of body and mind. In the creation of a human world, Prometheus attempts to restore the balance of mind and matter, heaven and earth, spirit and body, which the wars and feuds of the gods had upset. Prometheus gambles on humanity to do a better job of things than the gods had done. And so human beings are both blessed and burdened with the power to use the Promethean arts and sciences for good or ill.

As the giver of culture, Prometheus gave us fire—fire

to cook and refine our food, to dispel darkness and evil, to warm and brighten our dwellings, and to fuel the engines of civilization. But Promethean fire is far more than physical. It is the brilliant fire of the intellect and imagination, the burning fire of passion, the smoldering fire of envy, the raging fire of hate, and the infernal fire of destruction. Fire can enlighten, energize, delight, cleanse, purify, inflame, trick, cajole, create, transform, and consume.

Promethean fire is the foundation of science: Chemists and biologists deal constantly with the notions of combustion and oxidation—they even talk of life itself as "slow fire." Fire and heat are the basis of energy and its conservation. The heat engine and the laws of thermodynamics were central to nineteenth-century physics and culminated in the concept of entropy and the "heat death" of the universe. Twentieth-century science has been an all-out display of fireworks. The big bang is the great fireball from which the cosmos was born. An intense hidden fire of fission and fusion, buried in the heart of the atom, can give rise to nuclear energy and the conflagration of the bomb. Nuclear fire is also the fuel of the stars and our sun—it supplies the whole universe with heat and energy. Then there's vulcanism. The molten core of the earth is the forge in which metamorphic rock is formed, the source of volcanic lava, and the engine that moves the great tectonic plates of the continents. Technology—both good and bad—is utterly saturated in fire: smelting steel, burning coal and oil, combustion engines, napalm, incendiaries, bombs, rockets, even mass genocide of millions in cremation ovens.

What is the source of all this fire? Prometheus stole his flame from the fiery chariot of the sun and risked giving it to humans to use wisely. Today the fire and energy of the sun has become a potential source of creation or destruction as we struggle with solar energy, the fusion reactor, and the hydrogen bomb. Indeed, our modern struggle has become an allegory of good versus evil, meaning versus chaos, creation versus destruction. On the one hand, the slow, even controlled, use of sunlight, after it has been rarefied and tamed by the vast spaces between earth and sun, represents a gentle harnessing of the sun's chariot in the form of solar energy. The bomb, on the other hand, is a concentrated and rapid release of too much solar fire too quickly. An exploding hydrogen bomb, after all, imitates the hydrogen fusion occurring far away in the depths of the sun under extreme conditions, utterly different from those on earth.

RENEWING THROUGH MYTH

Of course, no one claims the bomb to be anything more than an irrational engine of destruction. The real competitor to solar power is the hoped-for fusion reactor—a slowed-down bomb, as it were. Here the conflict indeed takes on mythical proportions: the gentleness of Promethean heat versus the violence of a Frankensteinian explosion.

There is something basically harmonious and agreeable about capturing the sun's rays at the surface of the earth and storing or converting the radiant energy they contain. Most energy available on the earth today has been gained through some variant of this method. Photosynthesis, the biological fabrication of nutrients under the action of sunlight, is ultimately responsible for most of the food and fuel we consume.

The major exceptions to this are certain important natural edible substances, such as water and salt. (Despite their importance, incidentally, they have no nutritive value in themselves, and, in any event, they would essentially be unavailable to us without the action of sunlight.) There is also the natural radioactive material in the earth's crust, but this is a completely impractical (not to mention harmful) source of energy. The harvesting of the nearly limitless quantities of sunlight that naturally bathes the earth represents a gentle, harmonious, practical, and ecologically sound relationship between us, the earth, and our planetary environment.

It may be objected that advocating solar energy from what is largely an aesthetic and mythical point of view seems too subjective and idealistic. Furthermore, the argument seems to suggest that whatever is given in nature cannot be improved on and that technological progress is inherently destructive.

In the first place, to discount subjective and aesthetic arguments is to beg the whole question of scientific objectivity and authority. The preeminence of scientific thinking today is a consequence of complex historical events and psychological developments. The discrediting of subjective judgment was consciously built into scientific thought by Bacon, Descartes, and Galileo, all of whom advocated a materialistic and pragmatic approach to the natural world.

But the advantages and successes of science can no longer excuse its shortcomings and failures. The fact that something works is no guarantee that it is logical, valuable, ethical, or even useful. If we are ever to reevaluate and moderate science, we must stop appealing always to the same pragmatic and economic arguments that have gotten us into the many technological impasses we face today. As Owen Barfield suggests, we must place greater trust in human intuition and imagination as guides to our ethical choices.

As to the objection based on the idea of progress, it is a red herring. I do not claim that whatever nature does is right. Illness is natural, but very few people would oppose medicine. Yet, even the practice of medicine can be overdone, as is often the case in the extreme and costly life-preserving measures of modern medical technology. Nature must be neither slavishly imitated nor stubbornly opposed. Instead of these two extremes, nature can provide us with an aesthetic reference point—a guide to harmonious and meaningful progress. The best medical therapy works hand in hand with nature to nourish and heal. Even in combating illness, we recognize the aesthetic superiority of one form of therapy over another.

Photosynthesis, heat absorption by water and rock, ocean currents, wind circulation, sun baking, evaporative cooling, and other natural processes can guide and inspire us in finding ways of harvesting solar energy. We should seek and test methods for harvesting radiant energy that are not only efficient and economical but beautiful and harmonious, as well—methods that maintain the balance and order of nature. Solar heating panels and solar batteries have this potential, for they are benign and relatively unobtrusive surface collectors of energy, which, after all, is what leaves are.

Manufacturing solar panels can be done on a relatively small scale so that it remains an appropriate, economical, clean, and energy-efficient technology. Solar batteries demand a more advanced technology, and a more careful balance must be maintained between their ultimate utility and the natural resources required in their manufacture. But these details will evolve as solar technology improves, so long as aesthetic criteria are honored along with practical considerations. We need only follow the Promethean prescription of using fire and wit to benefit and enhance the human enterprise.

CHAPTER 20

Physics as Art

IF THE MYTHICAL AND METAPHORICAL nature of physics is a well-kept secret, its artistic qualities are more widely recognized. Many physicists speak of the deeply gratifying beauty inherent in such aesthetic characteristics of physics as simplicity, economy, elegance, symmetry, harmony, order, and resonance. Freeman Dyson goes so far as to claim that if a theory lacks beauty, it cannot be valid. Yet, all of this is a kind of lip service. In the last analysis, the acid test for any scientific theory is its empirical validity, and beauty be damned. Accuracy, not aesthetics, is the ultimate criterion for science.

If nothing else, science has had an influence on the arts of the late-nineteenth and twentieth centuries and is reflected in them. Concepts of modern physics are reflected in the atomizing effects of pointillism, in the variable fractured space of cubism, in the fluid space of surrealism, in the temporal imagery of futurism, in the random non-deterministic harmonies of atonal and serial music, in the nonrepresentational character of abstract expressionism, and in the loss of meaning of the theater of the absurd.

But from a wider and more philosophical point of view, we see more than the mere influence of physics on the arts. We see physics itself as a form of art. Much of the imagery and symbolism of physics is metaphorical and therefore constitutes a kind of visual and poetic creation. In addition to the visual and conceptual imagery that we find in physics, there are also narrative, thematic, literary, and dramatic elements: The big bang is a fabulous myth with a sensational opening, followed by a long evolutionary saga and a tragic ending. The struggle between conflicting scientific theories is like a courtroom drama. The very inspiration for certain discoveries—the benzene ring, for example—has come from dreams and archetypal symbols. Even such deeply held assumptions as reductionism and materialism are in the nature of literary themes.

THE ELECTRONICS CONNECTION

Looking at the equipment used in contemporary rock concerts and recording studios, you could easily imagine that it's all for some physics experiment or technical application. The modulation, control, recording, and amplification of music has been changed forever by modern electronic and computer technology.

Not only is music manipulated electronically, it is even created electronically. Electronic guitars and keyboards are essentially new instruments that musicians have made their own and which they use to perform music freely and idiomatically. The composition of music through the use of synthesizers and computers has become standard practice. Artists even use electronic technology to create "sound sculptures" and "sound environments" in such nonmusical settings as museums and office buildings.

The music industry is really a child of twentieth-century physics. For example, in less than a decade the compact disc, which combines the technology of digital electronics with that of the laser, has revolutionized the transcription, processing, and playback of recorded music.

Virtually every piece of equipment used in making and playing recorded music, from the studio microphone to the home loudspeaker, is based on the work of physicists and engineers.

PHYSICS IN ART

The arts have been profoundly influenced not only by the technology of physics but by the ideas and theories of physics as well. In the visual arts, especially, we see many suggestions of a cross-fertilization between new art forms and the concepts of physics. Such late-nineteenth-century pointillist painters as Georges Seurat and Paul Signac filled their canvases with tiny dots of color that, when viewed from the proper distance, give a synthesized impression to the eye of familiar colors and shapes. This technique reflects an atomistic point of view and also the idea that light and color can be reduced to a few simple elements. Similarly, impressionist painters Camille Pissarro and Claude Monet, and impressionist-influenced Maurice Utrillo, used short brush strokes and bright colors to imitate the physical effect of light on objects and to re-create the impression of light on the eye.

In the twentieth century, we find even stronger hints of modern physics in the visual arts. The cubist paintings of Picasso and Georges Braque portray a kind of "fractured" view of space, which represents many points of view simultaneously. This is quite suggestive of relativity theory, in which various observers describe space and time differently. Cubism, in fact, reflects the more general character of relativity, with its breakdown of the former rigid conception of space and time.

These distortions, as we have seen, are carried to a further extreme in general relativity, with its curved space-time. Such a fluid and curved notion of space is typically found in the surrealist paintings of Salvador Dalí, Joan Miró, and Yves Tanguy, with their stretching, twisting, and warping of plastic body surfaces.

Time, too, is represented in painting—for example, in the futurist work of Umberto Boccioni and Giacomo Balla. In Marcel Duchamp's famous *Nude Descending a Staircase*, time is indicated by a kind of "multiple exposure" of sequential images. Of course, such paintings also reflect the influence of motion pictures. But the explicit representation of time and motion in a single "still" frame is most suggestive of the new "democratic" treatment of time and space in relativity.

The philosophical implications of modern physics also find expression in twentieth-century painting. The works of Giorgio De Chirico picture bleak geometrical cityscapes and sterile interiors peopled by robots and partly dismembered bodies in aimless and ambiguous poses. The pictures of De Chirico and other metaphysical artists present an alien and hostile view of a random and meaningless reality, which is not difficult to associate with the chaotic chance fluctuations of the quantum world.

The paintings of René Magritte present enigmas that hint at the illusory and representational character of reality. In Magritte's *Les Promenades d'Euclide*, a painted canvas on an easel is set in front of a window, blocking off a portion of the outside scene of houses and streets. But the picture painted on the canvas appears to be an exact replica of the scene that must lie behind the canvas. The matching and registration of the painted and directly "viewed" scenes are nearly perfect, but not quite. It all leaves the viewer with the uneasy feeling that if the canvas were moved, it would leave an empty hole in the revealed scene. What is reality and what is representation? Magritte seems to ask us. In this day and age of a quantum theory that gives precise predictions but no depiction of reality (or even a guarantee that reality exists when we're not looking), Magritte's work seems all too poignant.

A more recent work, *Area Code*, by James Rosenquist, is a painting and construction that comments ironically on the confusion and complexity of modern electronic communication. In his larger-than-life work, Rosenquist depicts a huge phone cable with its colored, twisted wires exposed. The background medium is a sheet of silvered mylar, which is bent at right angles at each end so as to add reflections that contribute to the complexity of the whole view. Rosenquist's sly use of reflective materials accentuates the tangled quality of the phone lines and hints at the repetitive, reverberating character of the millions of crossed messages in the telephone network. The reflective foil also draws the viewer into the picture and makes her a part of the bewildering communication system. Finally, the reflections evoke the quality of art as illusion . . . or is it reality as illusion?

Then there is the random look and grainy quality of Jackson Pollock's work, which suggests the modern atomic view of matter, with its discrete quantized structure and random indeterminate motion. More generally, the characteristic mid-twentieth-century movements in the arts—abstract expressionism in painting, atonality in music, and the theater of the absurd—all seem to mirror the purposeless and meaningless view of existence that is implicit in quantum randomness and the purely materialistic philosophy of science.

PHYSICS AS ART

We see that art reflects both the outward appearances and the inner values that we associate with modern technology and physics. But is the reverse also true? Is there art concealed within the imagery and concepts of modern physics itself? Might we display works of physics not only in a science museum but also in an art museum?

Many of the visual models and graphic representations of twentieth-century physics have artistic merit in themselves. Examples are the geometrical representations in general relativity of space as a curved surface or of a black hole as a whirlpool in space-time; the lobed, leafed, and shell-like patterns of the probability distribution of an electron in an atom; the star and ringlike forms of a diffracted beam of electrons or a laser; the rainbow of an atomic spectrum; or the view of the planets, stars, and galaxies in the cosmos.

But it is not only the visual imagery of science that has artistic merit. There is also great beauty and aesthetics in simply contemplating the phenomena and concepts of modern physics: Einstein's notion of the relativity of simultaneity, time dilation, and length contraction have created a fabulous new image of space and time. His famous equation, $E = mc^2$, has revealed a dazzling and unexpected symmetry between matter and energy. This symmetry has been enlarged in quantum theory to include the notion of matter and antimatter, which motivates contemporary research in elementary-particle physics. In fact, aesthetic considerations—matters of symmetry, order, simplicity, economy, and completeness—play an important role in stimulating and guiding scientists in the choice and design of their research projects.

The predictions of quantum theory are weird and inconceivable, yet they are also delightful and ironic. Electron interference demonstrates the mind-boggling wave-particle duality of electrons, and the results of the Aspect experiments hint at Einstein's dreaded "spooky actions at a distance." The Bohr model and quantum theory not only provide an explanation of the atom and its spectra but also describe a unique and novel structure of atomic and molecular matter, which has a character and charm all its own. Only quantum theory is capable of explaining conductivity and superconductivity, which it does through a symmetric and ordered view of atomic structure that is worthy of a mighty work of architecture.

Modern ideas on the nucleus and its components have led to atomic energy and the bomb—awesome and inspiring, despite their great potential for destruction. The study of the nucleus has also introduced new ideas of symmetry and unity into physics, which are evident in the latest quark models and field theories. These new theories offer the hope of an ultimate grand unification of all the forces of nature, which would be the envy of the greatest artists.

All of these concepts, models, and theories of twentieth-century physics are, in one sense or another, representations of physical reality. But they are much more. They carry an aesthetic value as well, for ultimately they are creations of the human mind. They bear the stamp of uniqueness, originality, and imagination that characterizes a genuine work of art.

PHYSICS AS LITERATURE

The aesthetic qualities of physics go beyond purely visual and conceptual forms of art. They extend to literary and dramatic forms as well. To think of physics in terms of story and plot may seem even stranger than to associate it with the visual arts. But a new form of humanistic scholarship, called narratology, looks for narrative roots in many disciplines—not only in literature but also in anthropology, philosophy, history, and even in science.[1] The basic idea of narratology is simply that humans love to tell stories, and they do so implicitly and unconsciously in many areas outside of literature. There are narrative elements, for example, in historical accounts, in philosophical arguments, and even in such scientific theories as evolution.

By the same token, the idea of narratology can be applied to physics. Finding literary and dramatic qualities in physics shouldn't be too surprising. After all, whatever else physics may be, it's a human activity and a product of human culture.

When scientists talk about their own field, especially

---

1. M. Landau, "Human Evolution as Narrative," *American Scientist* (May–June 1984): 262.

when they give an overview or introduction, they often present a history of the subject or the story of their own involvement with it. This seems perfectly natural. The history of a field helps us understand its character and development, and the story of someone's participation personalizes and humanizes the subject for outsiders. But, according to narratology, our urge to tell stories is not just casual and incidental. The very human need for chronological and causal connections actually helps to define and make us what we are. It does the same for our work. Part of the logic, structure, and rationale of any field is based on narrative elements that are supplied by human beings and that guide the growth and goals of the field. Physics is no exception.

What else, after all, is the big bang if not a great epic story? Whether we view it as a creation myth or as a drama, it has all the elements of a good tale. There is a sensational opening—a cataclysmic explosion in which all matter and energy, even time and space themselves, originate. Then follows a long soap-operatic middle in which the particles, atoms, and elements come into being and embark on the heroic adventure of creating matter, stars, galaxies, and life. This turns out to be a long and difficult path with many false starts, detours, pitfalls, and dead ends—a virtual "pilgrim's progress" suspense story. Finally, there are a pair of tragic, if prolonged, endings. In one, the universe runs downhill, matter degenerates, protons decay, and everything gradually fizzles out in a chaotic heat death. In the other version, the whole pageant of creation and destruction plays itself out again and again in an endless series of mindless reruns.

Quantum theory, too, has its narrative elements. The long historical struggle between the wave and particle theory of light was resolved in the twentieth century through a synthesis of the conflicting ideas. We now have neither particles nor waves but "wavicles," which exhibit the properties of one of the two forms depending on the circumstances. We can interpret this debate through the traditional developmental pattern in philosophy of thesis, antithesis, and synthesis. Alternatively, we may view it as an example of conflict resolution between adversaries—a virtual courtroom drama.

Gerald Holton, a historian of science, has dealt extensively with the question of scientific narratology.[2] Holton argues that behind many scientific ideas there are subjective themes and beliefs that have no realistic justification. A case in point is that of Johannes Kepler, the great German astronomer and mathematician whom we discussed earlier. Kepler was the first person to describe correctly the planetary orbits as ellipses rather than the circles that everyone since Plato had believed in. Even Kepler himself resisted the idea of elliptical orbits for many years. Although an ellipse provided the best fit to the data, particularly to the orbit of Mars, Kepler initially could not give up Plato's perfect circular orbits. He had accepted Plato's thematic belief that circles are the ideal geometrical form and, therefore, that the heavenly planets must follow circular paths. Only after many years did Kepler reluctantly accept ellipses, and then only after he had convinced himself that the sun, located at the focus of an ellipse, plays the same central role as in Plato's circular orbits.

Another historian, Thomas Kuhn, argues that scientific theories come and go somewhat like fashions and styles in clothing.[3] Kuhn assigns the term *paradigm* to a scientific theory that is widely accepted during any given historical period. Classical physics and relativity, which replaced classical physics, are examples of Kuhn's paradigms. But the relativistic paradigm did not win out over its Newtonian counterpart simply by the force of the evidence. Many die-hard scientists clung to the old theory

---

2. See G. Holton, *Thematic Origins of Scientific Thought* (Cambridge: Harvard University Press, 1973).
3. See T. S. Kuhn, *The Structure of Scientific Revolutions, Second Edition* (Chicago: University of Chicago Press, 1970).

despite all proof to the contrary. It was only the passing of
the old guard and the coming to power of the Young Turks
that ushered in the new theory.

THE ACT OF CREATION

Literary and dramatic themes are found not only in the
texture and flavor of scientific theories but in their very
creation. Einstein, for example, completely reversed his
philosophy of science as a result of his awareness of the role
of his own imagination in developing the general theory of
relativity.

Early in life, Einstein had looked upon a scientific
theory as simply a logical generalization from the facts.
Later, after his inspiration for the general theory of relativity, he acknowledged the fundamental part played by the
human imagination in the creation of scientific theory. An
intuitive faith in the simplicity and beauty of natural law
had guided Einstein in his quest for a general theory and
led him ultimately to his unifying equations. For Einstein,
after general relativity, scientific theories were free creations of the human mind, subject to such aesthetic criteria
as simplicity, economy, and order. Only after their creation
are theories tested against empirical evidence, and even
then, Einstein had more faith in the beauty of a theory
than in the "facts."

Amazed by Einstein's confident reaction on receiving
the news from Eddington that the 1919 eclipse observation
had vindicated his general theory, one of Einstein's students asked Einstein how he would have reacted if the
theory had been proven wrong. "Then I would have been
sorry for the dear Lord—the theory is correct," was Einstein's famous response.[4]

There is also the case, already referred to, of Kekule,
who traced the inspiration for his discovery of the benzene

---

4. See C. M. Will, *Was Einstein Right?* (New York: Basic Books, 1986),
88.

ring to a dream in which he saw a snake circling around to
bite its own tail. This symbol is a classic archetype according to Jung—the *Uroboros*, which represents both the idea
of the many in the one and also the quality of self-genera-
tion. Not a bad choice for a symbol to unlock the mysteries
of organic chemistry.

THE FADED PAGEANT

The relationship between the human mind and physical
reality is a subtle one. Quantum theory has raised a serious
if ambiguous question as to the independence of mind and
matter. It tells us that an observer has a random and unpredictable effect on what is observed. I cannot observe an
electron without disturbing it. This suggests that physical
reality (whatever it is) cannot be independent of the human
mind that observes it. Our view of reality must of necessity
be colored by the aesthetic and metaphorical themes of our
consciousness.

This puzzling, problematic approach to nature is typical of modern physics, which has become highly abstract
and mathematical—not all that different really from the
imaginary creations of art. Matter, for example, is no
longer made of any stuff or substance. It has an idealized,
surrealistic character—a mere castle in the air. Electrons
are not hard little balls. They are "wavicles," and finding
them is a matter of chance, like ghosts in a Gothic tale.
Space and time form a curved four-dimensional continuum—a very abstract kind of geometry with a decidedly
science-fiction flavor. Black holes form bottomless pits in
the strange, warped space-time of relativity but appear
regularly in fantasy space movies. Quarks, the basic constituents of matter, are buried deep within the nucleus,
locked in a permanently unobservable condition. They are
more like a collection of mathematical ideas than the material particles of yesterday's physics. Or perhaps they are
princesses imprisoned in fairy-tale towers.

Electron waves, curved four-dimensional space-time, invisible black holes, concealed quarks, remote quasars, probabilistic atoms . . . are these elements of reality, figments of the imagination, or creations of an artist? Indeed, modern-day physics looks more and more like art, illusion, and show biz. It's hard to know what's real anymore. Reality may be the construction of a mathematician or of a stage designer. I think Shakespeare puts it best in *The Tempest* when Prospero, the supreme magician-artist, bids farewell to his magical powers of creation:

> Our revels now are ended. These our actors
> (As I foretold you) were all spirits, and
> Are melted into air, into thin air,
> And like the baseless fabric of this vision,
> The cloud-capp'd tow'rs, the gorgeous palaces,
> The solemn temples, the great globe itself,
> Yea, all which it inherit, shall dissolve,
> And like this insubstantial pageant faded
> Leave not a rack behind. We are such stuff
> As dreams are made on; and our little life
> Is rounded with a sleep.

CHAPTER 21

Science vs. the Humanities

TREATING SCIENTIFIC IDEAS AND THEORIES as absolute
objective truth, ignoring their origins in the human imagination, and absolving them from any need for a humanistic analysis and critique have all contributed to the destructive separation between science and the humanities, which
has become critical in the twentieth century. The dominance of science and technology in government and academia is already a fact of life. More funding and support is
generally available for research and development in science
than in the humanities. Scientific and technological professions command the most prestige and garner the greatest
rewards in our society. History, culture, and ethics are
constantly devalued in favor of scientific and technological
knowledge. Even in the area of religion, the church long
ago lost the battle for the human soul to science, which has
become not only our way of life and sole arbiter of truth but
our virtual state religion.

The transformation and rehumanization of science is
a major task facing our age. Science must be confronted on
fundamentally humanistic terms. Can we condone any

longer science's defiant stance outside of the liberal arts? Can we afford to maintain the separation of science from philosophy and religion? Can we continue to judge science apart from its ethical and aesthetic implications? Can we tolerate a science that categorically denies human meaning, value, and purpose?

THE LIGHT OF THE LAMPPOST

Despite the long-standing and pervasive practice in the West, there is nothing natural or essential about separating the humanities from the sciences. We have already explored the common origin that science and religion shared in the human quest to find the meaning and purpose of existence. In earlier times, this search was treated holistically. What we think of separately as spiritual and physical matters were formerly considered one unified area of knowledge. In mythology, for example, divine influences and interventions commonly determine matters on the earthly plane. In Platonic philosophy, there is an essential bond between the ideal realm and the physical plane. Indeed, the study and contemplation of things physical is supposed to enlighten human beings and lead them to the spiritual realm of Platonic Ideas.

With Aristotle, however, things began to change. Although there were still important connections between the divine celestial spheres and the sublunary realm, there was a growing emphasis on the knowledge of physical and biological phenomena on the earth. Why was this so?

In the effort to make sense of existence, human beings sought order and meaning in their environment—in the stars, elements, plants, and animals. But although the original motivation was to find a rationale for human existence, a gradual shift of emphasis took place. To explain the ways of the gods to man remains a vague and problematic task. It is subject to individual interpretation, inspiration, and revelation. There are no final answers. It is a frustrating and taxing quest with no easy rewards.

On the other hand, the study of the purely *material* aspects of natural phenomena, unencumbered by the effort to interpret their divine or spiritual meaning, is a less frustrating and less ambiguous task with more immediate rewards. Describing the chemistry and biology of a rose has turned out to be more definable, achievable, and practical than attempting to divine the cosmic plan behind the rose in the first place. And so Aristotle and many who came after him began to emphasize material studies over metaphysical matters.

It's like the ironic tale of the man who searches for his lost keys under a lamppost, not because he lost them there but because there's more light there to see by. What we are capable of doing most efficiently and effectively will often sway us and make us forget the more difficult task that we set out to accomplish in the first place. Scientific work is not easy. But it has certain appealing, gratifying, and rewarding characteristics that are extremely rare or entirely lacking in such fields as theology and philosophy—a sense of immediacy and verifiability, a level of consistency and reproducibility, a feeling of contact with reality, a history of practical achievements, a well-defined mathematical language of description and prediction, an explicit methodology of procedures and techniques, an inherent intelligibility and communicability, an evolving and cumulative sense of progress, an involvement with the affairs of societies and nations, and an unprecedented aura of prestige and authority.

But for all its brilliant traits, fabulous techniques, and shining achievements, science has not brought us one jot closer to fathoming the human condition and the mystery of existence. Science has greatly ameliorated life on earth and has given us vast power and control over nature. That no one can deny. But if it is our souls and the meaning of life we seek, then the light of the lamppost has illuminated nothing.

Medieval Humanism

In medieval times, before the dawning of the modern
scientific age in the Renaissance and the Age of Enlightenment, significant forms of holistic knowledge still prevailed. The now-discredited and so-called pseudosciences
of astrology and alchemy embodied a worldview of spirit
merged with matter. In astrology, we find an incongruous
and unacceptable blending of two fields—astronomy and
psychology—which in modern times clearly lie on opposite sides of the gulf that divides the sciences from the
humanities. But this comparison and rejection is facile and
misleading. Medieval astrology does not portray the empty
abstract space of modern astronomy, whose inanimate
planets and stars ray their influences down on a distant
earth by means of some causal physical process, such as a
force of gravity or electromagnetism. The astrological
world is a living realm, like the Platonic cosmos, in which
mind and matter are one, and the activities of the heavens
and the earth are synchronistic reflections of each other. It
is a world of symbolism, consciousness, and synchronism
rather than space-time, force fields, and causality. It represents the ancient human struggle, however misguided or
meager, to find some purpose and meaning in existence,
and it does this through a unified, rather than compart-
mentalized, approach to knowledge. Despite all its modern
exaggerations and distortions, medieval astrology presents
a blend of scientific and humanistic knowledge from which
we can still learn something today.

It is not simply the false predictions and irrational
methodology of astrology that turn so many modern scientists against it. Rather, it is the worldview of astrology that
implicitly denies a universe governed by causality, reductionism, and materialism. The rules and regulations of
science simply do not allow for a world in which matter
and consciousness, body and soul, and science and the
humanities are united. It isn't only the lack of a physical
explanation for astrological influences that scientists object

to; it's the denial of the need for any causal explanation at all! Astrology and astronomy are not alternative descriptions of the same reality. They portray completely different worlds. They are different metaphors. The criteria of one cannot be used to judge the other. They must both be judged by larger, more humanistic criteria, which neither can dictate.

THE GAP WIDENS

At the opening of the modern scientific era, the final cleavage of scientific from humanistic knowledge was accomplished, however inadvertently, by Nicolaus Copernicus. That Copernicus displaced humanity from its central role in the cosmos has become a modern cliché. What is less well recognized, and is yet more profound, is the Copernican elevation of mathematical description from its former role of merely "saving the appearances" of the phenomena to its modern authoritative status as an actual and objective description of reality.[1] No longer was geometry merely a go-between—a means of reconciling the appearances of the flux and decay of the earthly realm with the order and perfection of the celestial realm. Copernicus's heliocentric circles and Kepler's ellipses were not simply a hypothetical rationale for the erratic wanderings of the planets against the orderly motion of the stars. They were a description of the actual motions of the planets in the empty but physical space of the universe.

Copernicus and Kepler were not really aware of the fundamental nature of the change they initiated. Copernicus died a confirmed Aristotelian.[2] And Kepler, as we have seen, struggled tenaciously to maintain Plato's circles. But

---

1. See O. Barfield, *Saving the Appearances* (London: Faber & Faber, 1957), and E. Grant, *Physical Science in the Middle Ages* (Cambridge: Cambridge University Press, 1989), 87.
2. T. S. Kuhn, *The Copernican Revolution* (Cambridge: Harvard University Press, 1957).

the die was cast. The work of Galileo, Descartes, and Newton helped turn the tide of Western thinking. Galileo and Descartes insisted on a sharp separation between things mental and physical, and on the necessity and efficacy of mathematics for expressing the true laws of nature. The Cartesian world is a vast machine of matter and motion, obeying mathematical laws—a view that was brilliantly vindicated in Newtonian mechanics. Newton described the known universe mathematically, established the physicality of the celestial realm, and counseled against mere hypothesizing. Even in his publications, Newton was careful to separate his alchemical and biblical studies from his purely "scientific" work. For many years, Newton's deep, lifelong interest in esoteric and spiritual matters was the best-kept secret in science. But twentieth-century scholarship has changed all that.[3]

We can see this watershed in intellectual history reflected as well in the birth of early Renaissance perspective painting.[4] After the fifteenth century, it is the geometrical character of the relationship among things in space that becomes paramount. Spiritual, hierarchical, social, kinship, affinity, symbolic, mythic, and organic relationships become subsidiary and ultimately invalid by comparison.[5] Indeed, our very notion of space is first and foremost geometrical and mathematical. The central notion of causality in science is intimately tied to the geometrical conception of space in classical physics and relativity and to mathematical relationships in quantum theory. Ultimately, perspectival space became "official" space.

The rest, as they say, is history. The fabulous successes of physics in the nineteenth and twentieth centuries provided a dramatic contrast with the continuing frustrations

3. F. E. Manuel, *The Religion of Isaac Newton* (Oxford: Oxford University Press, 1974).

4. S. Y. Edgerton, *The Renaissance Rediscovery of Linear Perspective* (New York: Basic Books, 1975).

5. S. R. Bordo, *The Flight to Objectivity* (Albany, N.Y.: SUNY Press, 1987), 69.

of philosophy and the general decline of religion and theology. Science—and especially physics—became king and remains so to this day. The light of the lamppost has blinded us all.

MODERN DANGER SIGNS

In the modern era, the rift between the sciences and the humanities is all too apparent. The compartmentalization of knowledge permeates our academic and even our governmental institutions. At the University of Minnesota where I teach, the division was made official in the 1960s when the College of Science and the Liberal Arts was divided into a College of Liberal Arts and an Institute of Technology. Even after such a division, the departments of chemistry, physics, and mathematics did not wind up together with the liberal arts where they belong. Instead these departments chose to join the Institute of Technology with its primary emphasis on engineering, not because it was more appropriate but because that was where the money was. This pattern is hardly unique to Minnesota: it pervades the United States.

The problem is not simply the separation of the two areas. It is the dominance of one over the other. It is not merely a matter of intellectual equality (which, by the way, doesn't exist at many universities where the science faculty typically is better paid than the humanities faculty). Perhaps in government and industry, which have become so utterly dependent on modern science and technology, it is understandable that science is supported and financed far better than the arts and humanities. In academic institutions, which are devoted to intellectual equality and academic freedom, what possible justification can there be for the favoritism shown to the sciences? Hi-tech, big-buck, large-scale research in science, engineering, and medicine have become the bread-and-butter activities of all major academic institutions of "higher learning."

Even within the liberal arts, the dominant and best-supported fields are those that most successfully imitate physics. Typical are fields like political science, experimental psychology, and geography, which have become quantified and empirical and are broadly applied to the everyday affairs of government, politics, and business. "Weaker" fields like English and humanities itself must hobble along and try to get on the bandwagon.[6] English professors publish papers on the word count rather than on the literary attributes of Shakespearean plays. Humanists concern themselves with deconstruction and discourse as they analyze the literal meaning of texts rather than their inner sense.

All of this demonstrates the "scientization" of the humanities, while the humanization of the sciences is nowhere to be found. The evidence goes far beyond the boundaries of academia. In the area of religion, for example, the church long ago lost the battle over the human soul to science. Four hundred years ago, the Inquisition persecuted Galileo and burned Giordano Bruno at the stake because their teachings conflicted with Catholic dogma. This is a page in history that the church would be happy to expunge. And yet, despite its excesses and cruelty, the effort to squash science was no miscalculation. The church knew very well what it was doing. It knew who its enemies were, and it struggled to destroy them. In the long run, of course, it failed. Religion could not conquer science and neither can philosophy, sociology, history, or any other branch of the humanities. Any day now, we expect the coup de grace to be administered to our immortal soul when computers become bonafide thinkers and life is produced in the test tube.

When it comes to establishing "truth," which method is invariably selected? Why, the scientific method, of course. Again we see the complete surrender of the church to the authority of science in the canonization of saints and in the testing of religious artifacts (like the Shroud of

---

6. In 1991, the University of Minnesota voted to dissolve its Humanities Department.

Turin), which must pass through the rigors of exhaustive scientific inquiry. No one can deny the efficacy of science in its own theater of operations. The character of a miracle, however, cannot be judged by whether or not it complies with the known laws of science. It is the human and spiritual content of miraculous events that is paramount and to which science can contribute absolutely nothing. The rules of the game of science simply cannot be applied to religion, politics, and many of the affairs and experiences of everyday life (particularly the life of the mind, the heart, and the spirit). To insist on using science to arbitrate truth in such areas makes about as much sense as trying to apply the rules of chess to a game of Monopoly.

Even in the popular imagination, science tends to dominate. It isn't only that science and technology are pervasive on television and in film and in literature. More significantly, our very conceptions of the world are strongly colored by science. Who, for example, has had more influence on popular notions about reality, space, and time—Shakespeare or Newton? "The time is out of joint," means nothing to most people. But space and time travel and the big bang scenario are thoroughly integrated into modern parlance and imagery. The humanistic conceptions of reality embodied in literature, philosophy, and religion are not well known or understood, and when recognized, they are rarely taken seriously. Should such conceptions come into conflict with the prevailing scientific view, they are generally dismissed and rejected. Those who continue to hold views that contradict science are generally regarded as fanatic, kooky, irrational, or insane. Science is no longer merely a field of study—an "academic" discipline. In our culture, it has become a way of life and a system of belief. At its worst, science is a form of idolatry.

TAMING AND TRANSFORMING SCIENCE

The danger signals, destructive influences, and dominance of science over the humanities can no longer be discounted, trivialized, or ignored. Humanists cannot escape the effects of science by minimizing it or by omitting it from their school curricula and everyday concerns. Scientific literacy has reached a nadir in America, which further aggravates the problem. Ignorance is no excuse.

In fact, humanists in all fields, including the sciences, must take on the responsibility of reevaluating science and restricting it to an appropriate role. This can be done only by confronting science on humanistic grounds. In any effort to analyze and criticize science, the criteria cannot be dictated by science alone. If science is judged by its own rules, its humanistic faults and deficiencies will appear insignificant or irrelevant. Science is fundamentally a human activity, and its value and meaning in human affairs ultimately cannot be judged only by its many practical consequences, its considerable intellectual contributions to human understanding, and its great appeal and attractiveness as a human activity. It must also be subjected to questions and criteria as to its fundamental value and role in human culture and civilization—and judged accordingly.

Is it reasonable to accept and canonize a scientific worldview that fails to offer any meaning, value, or purpose for human existence? Are science and religion fundamentally on a collision course? If so, how can we resolve the conflict? Is science's much heralded separation from philosophy justifiable on humanistic grounds? Does the difficulty—even the insolubility—of certain fundamental human questions warrant our focusing on less profound but more tractable questions? Can the ideas and products of science be isolated from their aesthetic and ethical implications? How should aesthetic and ethical considerations affect scientific thought and research? What are the limitations and shortcomings of scientific theories? What are the basic assumptions used in science, where do they come from, and what are they based on? How has scientific thought evolved over the ages? Is the impermanence of scientific theories adequately publicized and recognized?

Does science always provide the best methods for determining truth?

If humanists will not ask and attempt to answer such questions, then who will?

"Humanistic" Scientists

Finally, we'll look briefly at three scientists who might serve as role models for a more humanistic approach to science—Johann Wolfgang von Goethe, Henry David Thoreau, and Barbara McClintock.

Goethe was the proverbial Renaissance man.[7] He lived in and helped to forge the Romantic Age, which was in part a reaction against the dominance of science. Nevertheless, he was a scientist as well as a critic, journalist, painter, theater manager, statesman, novelist, educator, playwright, poet, and natural philosopher. Goethe worked in both biology and physics. He developed a theory of the metamorphosis of plants in which the leaf was a kind of archetype for all other plant organs. He relied on both observation and intuition. He believed in a grand spiritual synthesis of the world and thus that imagination and intuition play a fundamental role in scientific research and study. He worried about the hazards of "purely" scientific thinking and sustained a quest for unity and continuity in nature.

His theory of color is largely discredited though still somewhat controversial today.[8] Goethe did not accept the Newtonian decomposition of light into component colors. He believed that light was fundamentally indivisible and that colors were merely the gradations of light and darkness. His physical ideas had some influence on Hans Christian Ørsted and on the development of the concept of the conservation of energy.

---

7. See N. Boyle, *Goethe—the Poet and the Age*, vol. I (Oxford: Clarendon Press, 1991).
8. F. Amrine, F. J. Zucker, and H. Wheeler, eds., *Goethe and the Sciences: A Reappraisal* (Dordrecht: D. Reidel, 1987).

Many contemporary biologists and physicists would not consider Goethe to be a true scientist. His methods do not jibe with the standard approaches of science, and he allowed his spiritual philosophy to influence and guide his work. But, of course, this judgment begs the whole question of scientific validity and the humanization of science. The real question is whether we might learn new ways to do and evaluate science by taking Goethe's work and methods more seriously.

Thoreau would be considered today as even less of a scientist than Goethe. He was a supreme writer and essayist as well as a surveyor, naturalist, inventor, businessman, and social critic. But he certainly considered himself to be a scientist.[9] His science, however, would have to be evaluated as a consummate example of humanism applied to nature. As a single instance, consider the famous "sandbank" passage from *Walden* in which Thoreau describes and reflects on the myriad provocative and grotesque forms taken by thawing sand and clay in the walls of a railroad gulley in early spring.[10] In this remarkable segment, Thoreau combines and integrates biology, geology, evolution, philosophy, religion, linguistics, semantics, art, and literature. Such a synthesis would be considered too speculative and imaginative to count as "pure" science. But still it must stand as a superb example of science interpreted humanly, philosophically, and spiritually.

Barbara McClintock is a scientist by anyone's definition. In fact, she is a Nobel laureate in biology, although the recognition of the importance of her work was delayed for many years by prejudice, incomprehension, and, of course, by the fact that she is a woman.[11] She did pioneer-

9. W. Rossi, *"The Laboratory of the Artist": Thoreau's Literary and Scientific Use of the Journal* (Ph.D. diss., University of Minnesota, 1986).

10. See H. D. Thoreau, *Henry David Thoreau* (New York: Library of America, 1985), 565–68.

11. E. F. Keller, *A Feeling for the Organism: The Life and Work of Barbara McClintock* (New York: Freeman, 1983).

ing and foundational work in genetics by establishing that genes could "migrate" and thus affect the laws of inheritance in unexpected ways. McClintock's work is based on entirely unconventional intuitive techniques. She "makes friends with genes." Only by befriending and developing an intimate relationship with each of the corn plants used in her research was she able to make her astonishing discoveries, which no one believed for forty years.

McClintock thinks that much scientific research is too focused on classes and numbers and too strongly motivated by preconceived and imposed answers to questions. She believes that one must be open and sensitive to the vast complexity of nature and to the uniqueness of the individual specimen. The individual is not the exception to a general class but rather the harbinger of some easily neglected clue that must be deeply respected and understood. This requires an acute and appreciative sensitivity to nature—an ability to "listen to the material." The traditional division of mind from nature is not only a misconception but also destructive of the proper techniques for scientific observation. Scientists today suffer from a lack of humility toward nature and its complexity. There is no guarantee of a central dogma into which everything will fit.

The values and attitudes exemplified by Goethe, Thoreau, and McClintock would take us a long way toward the humanization of science.

CHAPTER 22

Physics Education

DURING THE 1980s, SEVERAL STUDIES revealed the critical state of science education in America and the need for reform. Part of the response to these studies has been the effort to "beef up" introductory physics courses in high school and college by making them more mathematical and rigorous. While this may improve professional science and engineering training, it will not solve the grave problem of general scientific illiteracy. Laypersons do not need physics problem-solving skills but rather an understanding of the meaning and significance of physics. Technical education must be supplemented by a broad base of conceptual qualitative science education, which will help motivate more students to pursue careers in science. It will also produce a better-informed citizenry and electorate, helping laypersons understand and judge the profound influence science has had on modern life.

All of this will require a humanization of the physics curriculum. This means more than simply showing physics as the work of human beings and placing it in a historical perspective (as important as these are). It means ex-

351

ploring the essential connections between physics and human culture. We need to emphasize the human and personal element in our science classrooms by discussing the lives of scientists, as well as their motivations and inspirations. Physics must be taught from an integrative and cross-disciplinary point of view by exploring the complex connections between physics and social, political, philosophical, religious, and artistic matters. Finally, we must take a critical approach to science—learn to evaluate it from a broad humanistic and philosophical perspective—so that science can contribute to the spiritual as well as the material needs of humanity.

THE RELEVANCE OF PHYSICS

In the late 1960s and early 1970s, a wave of curricular reform swept through the colleges and universities of the United States. New course content and new methods of teaching and learning were explored in every possible setting from the traditional classroom to newly created experimental colleges. One of the bywords of this reform movement was "relevance." It was applied as a litmus test to every suggestion for change, and indeed to all existing courses and curricula. Unless the relevance of a subject or method could be established, it had no place in the education of the future.

Nearly a quarter century has passed since the heyday of counterculture reform, and "relevance" has gone out of style. Certainly it was both abused and overused during the reform movement. It cannot be the sole criterion for judging what we teach and learn. But relevance remains relevant, whether it's in or out of fashion. Physics is more relevant than ever as we approach the twenty-first century. The reasons for the relevance of physics, however, go far beyond the usual arguments given for studying it.

Of course it would be nice if everyone could understand how television, computers, and nuclear reactors work, but few people are willing to devote the long years of study necessary to master the intricacies of electronic and nuclear technology. For that matter, few people ever take the trouble to understand the rudiments of an automobile engine or even a bicycle. Much more is at stake than mere curiosity about the hundreds of technical devices in our lives, or even the ability to use and repair them with confidence. For the lasers, microwaves, stereos, VCRs, television sets, computers, and reactors are not simply there for us to use when they please us and to ignore when they don't. They have become pervasive in all forms of human activity and affairs. They influence and control us whether or not we are aware of them and understand them. Indeed, some degree of scientific understanding is necessary for self-defense, if nothing else.

Even if the effects of science and technology can be minimized in our personal lives, it is a practical impossibility to reduce these effects on the broad social, economic, and political level. The affairs of nations in the twentieth century have been profoundly influenced by the economic and political consequences of science and technology, to which physics has made no minor contribution (for better or for worse). All the people of the earth have been affected by the machines and technologies owned or controlled by governments—atomic weapons, modern electronic warfare, broadcasting and telecommunications, jet travel, miniaturized data processing and storage, medical technology, and space travel. And, of course, there are vast and powerful industries based on the technology of physics. It is not merely a matter of curiosity or convenience to have some glimmer of the fundamentals behind these powerful forces in our lives.

How can we possibly participate in government affairs or vote intelligently without some background in physics? So many questions today demand public participation and the voicing of intelligent opinions. Can nuclear reactors be made safe? How can nuclear wastes best be disposed of or stored safely away? Does the star-wars defense system make any sense? What is the danger of radon in the home, and should the government take action because of the danger? Will the fusion reactor ever be a practical source of energy, and how much government support does it deserve? Can industrial pollution be controlled, and how much of a long-term threat is it to the environment and life on earth? I would go so far as to say that meaningful and effective democratic governance today is impossible without an understanding of science generally and of physics in particular.

Yet even these reflections on the need for understanding physics do not reveal the depths of the relevance of physics in today's world. For there is something far more important than understanding how devices work and how physics influences us through government and industry. The concepts, theories, and assumptions of physics have profoundly and unconsciously affected our view of the world and our role within it. They have determined how we conceive of the objects and matter that fill our world and of the very space and time in which they exist. They have dominated our thinking about how the world and we have come about. They have provided the rationale, or the lack of it, for our very existence and being. They have, in short, given us the mental scaffolding that supports all our beliefs about the character, meaning, purpose, value, and essence of human life. This great and influential body of theories and assumptions has been acclaimed for the most part as the one and only truth about the world—or, at the very least, as the only legitimate approach to that truth.

This, in my opinion, constitutes the greatest relevance of physics in today's world and the most important reason for studying it. It is, in fact, my rationale for writing this book, with its emphasis on the meaning and implications, as well as the ideas, of modern physics. I believe that physics education—especially at the popular level—is even more important today than in the past if we are ever to appreciate the unconscious influence of physics on our lives and destinies. Therefore, I close this book with a discussion of physics education, because it is only through education that we can hope to revaluate science and rededicate it to the fulfillment of long-neglected humanistic goals.

THE PROBLEM OF PHYSICS EDUCATION

Why do so few high school students study physics, when it is the basis of all fields of science and technology today and offers knowledge essential for any educated person? Only 20 percent of American high school graduates have taken physics. The other 80 percent are inadequately prepared to be citizens and members in today's highly technological society. To all intents and purposes, they are scientifically illiterate. Why is this so?

We have already discussed the fears concerning physics that intimidate and alienate many students. The threat of a destructive technology, the folly of the quantitative, and the picture of a bleak, meaningless existence are dehumanizing and frightening aspects of the scientific worldview. Physics is viewed by many as a cold, abstract subject, antithetical to the world of feelings and emotions and oblivious to the psychological and spiritual needs of humanity. In addition, physics has gained a "bad reputation" in high school and is often avoided by students who are afraid of taking on a tough subject and of having to face up to their math anxieties. Unfortunately, some of this comes about because of antifeminist and anti-intellectual biases in American society. Despite decades of women's liberation, the percentage of women in physics has increased very little. This eliminates half of the potential population of physicists. While high school students earn "letters" and gain lots of peer admiration for athletic achievement, they get precious little credit or support for studying Latin, philosophy, or physics.

These problems are only aggravated by the typical and sporadic response to the "crisis" of American science education. In the 1950s and again in the 1980s, the poor and deteriorating quality of American science education became apparent on a national scale. Secondary science education, which includes physics education as a basic component, was singled out for the weakness of its curriculum, the inadequate training and poor morale of its teachers, and the low level of student motivation and interest in science.[1] These deficiencies have led to poor preparation for high school students seeking careers in science and engineering and have alarmed the academic and professional physics and engineering communities, which typically respond by advocating a toughening up of the curriculum. Calculus-based and advanced-placement physics are promoted at the high school level, and college entrance requirements are raised to weed out the weak students.

Whether or not such "solutions" actually improve the training and increase the numbers of future scientists and engineers is questionable. One thing it surely does is further alienate the large segment of the population that is not interested in professional training in science. Unlike students bound for careers in science and engineering, nonscience students do not need a working knowledge of physics and its problem-solving techniques. What they do need, however, is to be familiar with the broad ideas of physics, its history and philosophy, its social and political implications, and the profound effect it has had on all aspects of modern life. They need to understand how physics relates to and interacts with other areas of knowledge and interest in the arts, humanities, and social sciences. Above all, they need to be exposed to physics as a human and creative endeavor.

A Humanistic Approach

What is needed is a conceptual humanistic approach, both in traditional science courses and also in innovative new

---

1. See, for example, *A Nation at Risk* (Washington, D.C.: U.S. Department of Education, 1983) and *The Crisis in High School Physics Education*, special issue of *Physics Today* (September 1983).

courses for all students that integrate science with the humanities. In the study of cosmology, for example, the human aspects of the evolving unity between the big bang and the subatomic world must be emphasized. This unlikely marriage represents not only a brilliant synthesis of the macro- and microrealms but also a kind of culmination of humanity's legendary quest for an all-encompassing unity. Rather than presenting today's powerful telescopes and accelerators simply as remarkable technological tools of research, the mythical and symbolic role of these instruments in dealing with puzzles that have intrigued humanity for ages should be underscored. Where have we come from? What is the nature of reality? Is everything made of the same "stuff"? Is the world fundamentally a unity or a diversity? Such questions are traditionally studied in philosophy, anthropology, religion, history, and literature. It would be very revealing and meaningful for students to recognize that physics, too, sheds light on these deep human questions but that it offers no final answers to them.

It is ironic that popular books on physics have increasingly appeared on the bestseller lists at the same time that registration in science courses has been decreasing. Do publishers know something that science teachers don't? There is more to learning physics than equations, practical applications, abstract theories, lab experiments, hands-on experience, and problem solving. These are important, more so for formal training in science and less so for popular conceptual courses aimed at improving science literacy and qualitative understanding.

Furthermore, the emphasis in conceptual physics should be on the modern era, which is both more interesting and relevant in today's world. The concepts of classical physics, which are necessary as a background to the physics of the twentieth century, can be developed in context as needed, just as I have attempted to do in this book. The belief that formal training in classical physics and mathematics is an essential prerequisite to the study of modern physics may be justified for physics majors and engineers.

Such a background, however, is hardly necessary for the qualitative and conceptual understanding of nonscience students and the general public. Often the traditional order of presenting physics is more a matter of habit and convenience for the instructor than a necessity for the student. It is pure arrogance and hubris to claim that there is only one way to learn physics—the traditional path through Newtonian physics and higher mathematics.

The advocacy of a humanistic approach to science education in no way eliminates or diminishes the need to improve courses for training future scientists and engineers (although a little humanization would help to add enrichment and motivation in traditional courses). The two approaches should be parallel. Humanistic physics for nonscience students in school and for the lay public addresses a separate and equally urgent need in our society—to help overcome the general level of science illiteracy in America. Our real need is not only for more scientists but for more people who can think intelligently about science.

If future generations are not educated to appreciate the role, meaning, importance, and limitations of science, then it will become more and more difficult for Americans to make informed personal and political decisions and to maintain a democratic system. What's more, without a scientifically knowledgeable public, support for science will always be left to the experts, with their biases, and to the vagaries of political expediency. Finally, placing more emphasis on conceptual and popular science education ultimately may interest more young people in science and motivate them to enter the field. There's more than one way to skin a cat.

The humanistic approach not only involves presenting physics as the work of human beings but also as a subject integral to the liberal arts and humanities, because of the light it can shed on deep and long-standing humanistic problems and issues. Usually, a humanistic approach in science education implies the inclusion of some historical and biographical information. While this is useful and important, it does not begin to address the real conflict between the scientific and humanistic points of view, which continues to plague our society despite much lip service to the contrary.

Furthermore, placing physics squarely among the liberal arts means that physicists would have to deal with historians, philosophers, artists, sociologists, and even theologians on a more equal basis. A philosopher or art historian might offer a thoughtful criticism of physics, which could neither be dismissed as irrelevant nor ruled out as inapplicable to science. The liberal arts generally could benefit from such interdisciplinary exchanges and critiques. Most academic disciplines have become too specialized and ingrown for their own health. Fresh points of view and criticism could provide a much-needed breath of fresh air.

Physicists ought to be especially open to suggestions and reflections from friends and colleagues in other fields. Forces of unreason and irrationality are not far below the surface in twentieth-century civilization. Religious fanatics and proponents of doctrinaire and simplistic cults attract much attention and amass followings. The public, long deprived of any generally shared and accepted meaning and rationale for life, is hungry for some kind of spiritual or religious nourishment. This need cannot be ignored indefinitely or attributed simply to wishful thinking, anthropomorphism, and an unwillingness to "face reality." All of these trends, which we ignore at our own peril, contribute to a certain antagonism toward science.

By working hand in hand with sympathetic people in the humanities and the lay public, it may be possible to find new approaches and values for physics that will refresh and revitalize it. It must be a cooperative effort. Neither scientists nor humanists can do it in isolation from each other. The humanists need the scientists to explain the inner meanings of science and to interpret its theories. The scientists need the humanists to adapt new values, viewpoints, and critiques of science. If members of the scientific community cannot muster a little humility and participate constructively in a revaluation and aggiornamento of science, then irrepressible and irrational forces will continue to chip away at science and erode its foundation, or worse—destroy it through some catastrophic extremist revolution. For all its beauty and utility, a physics that ignores and rejects its own human character and the humanity of its creators and users cannot survive among human beings. Ultimately, physics does not belong to the physicists but to all humanity.

Index

361